Unwanted Bread

Stjordalen Lutheran Church near Reynolds in Grand Forks County is bathed in the rich color and deep shadows of an evening harvest sunset.

Unwanted Bread

The challenge of farming and ranching

SHELDON GREEN   JAMES COOMBER

Photography by Sheldon Green

■ INSTITUTE FOR REGIONAL STUDIES ■

Institute for Regional Studies
North Dakota State University
Fargo, North Dakota 58105

Copyright © 2000

Library of Congress Catalog
Card Number 00-108433

ISBN 0-911042-55-5

First Printing November 2000

Printed in Canada

Design by Sheldon Green

This book was also made possible by grants from:

The North Dakota Council on the Humanities
An affiliate of the National Endowment for the Humanities
Bismarck, North Dakota

Concordia College
Moorhead, Minnesota

The Cooperative Foundation
St. Paul, Minnesota

Cenex-Harvest States Foundation
St. Paul, Minnesota

The North Dakota Farmers Union
Jamestown, North Dakota

The Julius and Bertha Orth Foundation
Beulah, North Dakota

Gunlogson Foundation
Cavalier, North Dakota

Acknowledgements

One way to understand what is happening in farming today is to listen to those who are involved in it. These stories provide a range of perspectives – not only illustrating how complex North Dakota farm problems are but also demonstrating ways in which rural people might meet the challenges. Nearly everyone agrees farming is changing drastically today, and change will likely continue for years to come. Whatever our farmers and ranchers decide to do will profoundly affect life on the northern Great Plains.

Many people have helped make this book possible. We especially want to thank former North Dakota Governor George Sinner for his insights, guidance, encouragement and generous gift of time. The late Mel Maier and present administrator Neal Fisher of the North Dakota Wheat Commission were most helpful. Linda Liebert Hall of the U.S. Small Business Administration and Mikkel Pates, former farm reporter for The Forum of Fargo-Moorhead, now at Agweek in Grand Forks, helped us identify trends and interview possibilities. Demographer Richard Rathge at North Dakota State University was helpful with statistics and population trends. Dr. Tom Riley, director of the Institute for Regional Studies at NDSU, was enthusiastic about this project from the beginning and offered sound advice and vital financial support. Our colleagues at Concordia, particularly Linda Brown and Clyde Allen, responded with timely encouragement and crucial financial help. Hiram Drache offered his expertise on the history of farming, and Andrea Hunter Halgrimson of The Forum helped in locating news stories. As always, Ev Albers of the North Dakota Humanities Council was a willing listener and source of valuable advice for research and funding possibilities.

Various family, friends and colleagues helped with graphics, advice on style and with editing our manuscript. We wish to thank Karla Mickelson, Scott Olsen, Mark Strand, Kevin Carvell, Doug Carlson, Jon Suomala, Kelly Tucker, and Eleanor and Sarah Coomber. Leah Bendavid-Val and Sam Abell, both of the National Geographic Society and leaders of the book publishing session at the 1999 Santa Fe Workshops, provided generous technical and professional advice and solid encouragement.

Just as finding investment capital for value-added ventures is difficult in North Dakota, locating funding for publishing is an exhausting task. Several organizations provided the financial support for research, travel and publishing. We thank the North Dakota Institute for Regional Studies; Concordia College-Moorhead; North Dakota Council on the Humanities; North Dakota Farmers Union; Cenex Harvest States Foundation; Cooperative Foundation; the Julius and Bertha Orth Foundation of Mercer County; and the Gunlogson Foundation of Cavalier.

This book would not be possible without the generous help of the farm families and agribusiness people we interviewed. They welcomed us into their homes and offices and answered our questions with patience and remarkable perception. If this book provides anyone with new insights and perspective, it is because of them. ∎

JAMES COOMBER is a professor of English and chair of the English Department at Concordia College, where he has taught since 1966. He received his Ph.D. from the University of Wisconsin at Madison and for many years chaired the Concordia Conference on Reading and Writing. Coomber is co-author with Howard Peet of the Wordskills vocabulary-spelling textbook series, as well as other publications in the teaching of English.

SHELDON GREEN is the senior writer in the Office of Communications at Concordia College. He has been the editor of the Hazen Star weekly newspaper and North Dakota *Horizons* magazine and helped edit, design and photograph the five-volume North Dakota Centennial Book series. He is a graduate of the University of North Dakota.

Grain elevators stand alone under the vast western sky south of Beach in Golden Valley County.

Preface

In 1995-1996 while we were traveling throughout North Dakota working on our previous book, "Magnificent Churches on the Prairie," we noticed distinct changes in the landscape. Occupied farms were becoming farther apart. Small towns were getting smaller and seemed to be populated mainly by aging people. Rural churches were closed and boarded up. Highways were fairly empty of traffic except for an occasional UPS or Schwan's delivery truck.

When we mentioned these observations to friends and colleagues back in Fargo, we were often told, "That's just the economy washing out the weak operators. It happens all the time." But we didn't think so. Something more fundamental, more threatening was happening right before our eyes. The face of agriculture – our primary economic engine – was changing. It was changing fast enough and significantly enough to be noticeable. After studying data provided by the U.S. Census Data Center at North Dakota State University, it became clear that North Dakota is shifting away from being a rural state of family-sized farms. It also appeared these shifts were being either unseen or ignored by a good number of people.

Farming is a complex occupation. What works for one operator may not work for another. Farm talk can also be confusing: deficiency payment, set aside, target loan rate, safety net, Freedom to Farm, GATT, PIK, CRP and on and on. A knowledgeable discussion on farm issues is challenging at best, and downright confusing to the uninitiated.

So how could we get the point across that our farmers are struggling?

Maybe by sharing a cup of coffee. That's right. If we could sit down with a farmer over coffee and listen to his or her story, we'd begin to understand the challenge farmers are facing today. And if we could hear stories from many different kinds of farmers, a more complete picture would emerge.

So we began studying the data to spot trends, then talked with experts at the North Dakota Wheat Commission, Farmers Union, Farm Bureau and so forth. We then assembled a list of farmers who represented the various trends in agriculture today. We found people who were getting out, staying in, diversifying, raising bison, planting specialty crops or attempting to start value-added cooperatives. During 1999 and 2000 we interviewed fifty farmers, ranchers and others specifically tied to agriculture to hear their stories. This book is the result of those interviews, most boiled down from hours of taped conversations.

Here, in their own words, North Dakota producers discuss the frustrations or opportunities they are experiencing in agriculture today. We are, in effect, offering you a table at your favorite coffee shop to hear these conversations. Former Governors George Sinner and Ed Schafer add their insights to the conditions in rural North Dakota, and we weigh-in with our own perspectives to what is being said on these pages.

From our conversations and research, we believe this is a time of significant change in agriculture in North Dakota and throughout the Great Plains. Potential solutions, like the problems, are complex and we make no recommendations, although many of the people we interviewed have specific ideas for helping those remaining in agriculture.

The people in this book were incredibly generous with their time. We feel privileged to have met them and to have made friends with them. ∎

Contents

Essays on agriculture

Fifty North Dakota farmers and ranchers

Val Farmer *Fargo*	12	Dr. Michelle Merwin *Hebron*	46	Michael McMullen *Fargo*	80
Jon Norstog *Fargo*	14	Gary Greff *Regent*	48	Robert Carlson *Jamestown*	82
Sharold Geist *Hazen*	16	Sonia Meehl *Crete*	50	Deb Lundgren *Kulm*	84
Rob & Jeri Dobrowski *Golva*	18	Jon Wiltse *Lisbon*	52	Scott Steffes *Fargo*	86
Frank & Deborah Popper *Highland Park, NJ*	20	Fred Kirschenmann *Windsor*	54	Father Roger Synek *Center*	88
Cindy & Kenny Gross *Edinburg*	22	Dennis Kubischta *Hope*	56	Albert Two Bears *Cannon Ball*	90
Bill Patrie *Mandan*	24	Mike Degn *Sidney, MT*	58	Gary Hanson *Grafton*	92
Doc Throlson *New Rockford*	26	Warren & Mara Solberg *Wild Rice*	60	John Fiedler *Mott*	94
Jim Harmon *Carrington*	28	Mark Willows *Binford*	62	Marilyn Hudson *Parshall*	96
Dan Wipf *Edgeley*	30	Bill Lowman *Sentinel Butte*	64	Mike McDougal *Rolla*	98
Clarence Johnsrud *Williston*	32	Dennis Sexhus *New Rockford*	66	Ed & Bunky Nistler *Beach*	100
Bonnie Woodworth *Halliday*	34	Lyle McLain *Mohall*	68	Mike Warner *Hillsboro*	102
Joe Satrom *Bismarck*	36	Rev. Dan Paulson *Alexander*	70	Dr. Gerald Sailer *Hettinger*	104
Doris Goettle *Donnybrook*	38	John & Chris Skogen *Epping*	72	Herlinda Martinez *Hamilton*	106
Rich Haugeberg *Max*	40	Norman Weckerly *Hurdsfield*	74	Dean Meyer *Dickinson*	108
Keith Bjerke *Bismarck*	42	Al Ulmer *LaMoure*	76	Chuck Suchy *Mandan*	110
Jim Weinreis *Sentinel Butte*	44	Tom & Joyce Scherr *Zeeland*	78		

To succeed in today's economy, farmers can no longer expect to be successful by merely delivering their grain to the local elevator and collecting a check.

America is losing its farmers

George A. Sinner
Governor of North Dakota
1985 - 1993

For farmers who are not part of the food processing and marketing industries the end appears to be near at hand. The socio-political consequences are not yet clear, but history does not have a good report on the amassing of productive land into the hands of bureaucracies, neither public nor private.

The human frustration, pain and despair are omni-present in rural America, as these interviews bear witness. For those who have had the opportunity and who have moved courageously into the food chain, there is promise, but for those who have not, the future is dim indeed. Nothing short of food shortage and panic among consumers will save them.

The stark reality is that America is losing its farmers at an incredible rate. The preeminent hope of almost all early immigrants . . . the hope of escaping the tyranny of the landed aristocracies of Europe . . . is methodically being dashed by the mindless forces of economics, profit taking and the disinterest of the best fed people in history.

In most cases, we are well beyond the question of "economies of scale" in farm bankruptcies. Most farm commodity prices are well below the "lowest world cost of production" and even greater size gives little hope of survival to farm producers.

American producers lost their markets in the 1980s when the dollar was allowed to become so bloated that even our own citizens could not afford our products when compared to those of other countries. Plains farmers were hammered by a drought of historic proportions at the same time. What had been a slow demise of farmer liquidity turned into a nightmare of bankruptcies. North Dakota lost nearly twenty percent of its farmers in the ten years of the 1980's and the 1990's have continued that pace.

Bloated currency and incredibly high real interest rates have teamed up with the vagaries of drought, flood and crop disease to doom the hope of sustaining individual ownership of productive land in the United States. During the last decade of this century it has been "just in time" management practices that have caused the greatest damage. This systematic reduction of inventories by all processors was a derivative of high real interest rates.

But farmers have had nowhere to go. The crops and livestock grew when they grew. But once the pipeline was full, no one wanted the stuff and the farmers' own high real interest rates had to be paid. Their only option, year in and year out, has been to sell their commodities at fire-sale prices. Ownership of a part of the food industry, where the profit margins have been much more reliable, has been the only salvation. But we have done a very poor job of getting farmers into that ownership.

And so, you will see here some very sad stories . . . told by intelligent, well-educated, patient people; most of whom, amazingly, still cling to hope. You will also meet some very creative, resilient people who have found a way to stay in business on the farm.

For my part, I pray that those leaders of the giant food industries who, I believe, really do not want to farm, will find it in everyone's best interests to make room for direct, farmer investment and participation in the food industry. I fear that, unless there is a friendly attitude toward this change, there is more than enough monopolistic power to wipe out any but the most carefully designed farmer efforts to survive. ■

Val Farmer
Clinical Psychologist ■ Fargo

Dr. Val Farmer specializes in rural mental health. He is nationally known for his weekly syndicated newspaper column that originates in the Fargo Forum, and his frequent appearances on radio and television programs to discuss how the farm crisis is affecting rural people.

"I see a lot of people wearing out in farming. They are just staying even – staying on the treadmill. The common way of putting it is 'it's no fun any more.' They understand it's not working and they feel they would like to get out.

"The single-minded, workaholic, perfectionist, hard-driving person who has put his life into his work is usually the one who's going to hang on. This farmer hasn't explored enough of life to know there are alternatives out there. So it's an all-or-nothing proposition. These farmers dearly love what they do and farming gives a lot of meaning to their lives. They are used to being in business for themselves and they see working for someone else as a form of slavery. There is a lot of fear about life away from their sheltered, rural community.

"On the other hand, the people who are likely to make good adjustments are those who have been college educated or perhaps gone to technical school. They have some confidence that they could succeed in some other part of society. They may have lived off the farm and have experienced a bit of suburban life and know it isn't that bad. They feel they have choices in life. Farming is one way to live a successful life, but it's not the only way. They are not as wedded to the rural ethic as being pure and virtuous and ennobling with the rest of the world being somehow not as worthy.

"There are legitimate concerns about leaving farming. Rural people have life-long ties with family and friends and community. Many are willing to be underemployed or fight to stay in agriculture simply because they don't want to disrupt the ties they have. They haven't experienced starting over in life, making new friends and building new support systems. They just can't imagine what it's like to go somewhere where they don't know anybody, where they have no history with people.

"The declining population in rural areas is affecting folks. There is depression that goes with missing people who have left and the friendship patterns that have been disrupted. There are fewer volunteers in the community, which means the remaining people are called on to do more and more. There is no surge of enthusiasm or optimism. Instead there is a sense of loss and decline that affects the general mentality of the people who are left.

"Another factor is when a farmer senses the agricultural economy is out of synch with the mainstream economy. When people are paid once or twice a year with uncertain markets it seems as though outside forces are controlling their lives and determining their happiness. It's easy to become angry when they feel they are not being treated fairly. They are working hard and producing food for the world, but the rewards just aren't there. When they can't pay their bills, it's easy to become angry and project their problems elsewhere.

"There are people who get together and complain – just because that is the way they talk when they're together. They feel a part of the group if they complain a lot. So there is an acceptance level of complaining that is normal. But there are some for whom the complaining becomes their whole worldview. They begin thinking they are victims of a great conspiracy. This is what I call victim thinking. It has a very powerful grip on the way some farmers think about everything. They feel helpless and powerless. They can become more emotionally reactive to problems because they feel they are being picked on all the time.

"The people most likely to come in to talk with a counselor are women in stress. Some are assertive enough to get their spouses to come along. Sometimes people just want to hear that it is okay to get out of farming. They know what it's doing to them. They are working way too hard; their relationships are suffering; their family is suffering; the rewards aren't there; the pressure of debts piling up is incredible. They don't have normal lives. They don't have vacations or weekends off. It's just a hard life. So they feel relieved when they make the decision. They just needed someone to encourage them.

"The keys to successfully adapting to what's going on today in agriculture is a combination of flexibility, good communication, support, religious faith, sense of humor and attitude. People need to be able to move to Plan B or C when Plan A fails.

"One of the dual messages is for people to be creative and use all of their management skills to get themselves out of a crisis. Hang in there. Be persistent and dedicated. Use the usual formulas that count for success in farming. The second message is this: there is life after farming. A lot of people have made the adjustment. It may take a little while, but things will work out. The old family farm lifestyle didn't turn out the way it was supposed to, but their new lifestyle may actually be much better for them in terms of relationships, more time and less stress.

"The era of the independent farmer is over. To survive, farmers need to share, perhaps get into some cooperative arrangements and work together. Or else they need to get bigger, and that is a treadmill that never ends." ■

On a rainy spring Saturday in the Red River Valley, Val Farmer attends an auction for a couple retiring from farming. Dr. Farmer counsels people about leaving the land in his practice specializing in rural mental health issues.

From an office in downtown Fargo, Jon Norstog manages more than 250,000 acres of farmland, and his portfolio is growing every year. For farmers wanting to stay in agriculture, he advocates tighter management procedures to reduce input costs.

Jon Norstog
Farm and Ranch Management ▪ Fargo

"I don't see a lot of optimism out there. I have one farmer I rent 2,000-acres to who is getting out this year. Another farmer has a nice diversified operation in the Red River Valley – potatoes, sugar beets, soybeans and wheat – who is also getting out. He's had enough. He doesn't want to lose any more equity in the farm. But I also have guys who are paying cash rents up front who are doing fine. It's hard to paint a broad picture of what's happening when you deal with so many individuals.

"As a farm manager, I'm viewed as an intermediary, not a member of the family or an outside threat. This way siblings can look each other in the eye at family picnics rather than worrying or fighting over what's happening on the farm.

"The sad part of farming today is people are moving off the land. I'm seeing people who own land living outside their communities or outside the state. Keeping local ties is important to some people who look at it as 'grandpa's farm' or 'mom and dad's farm.' But there are others who see cash renting only as an investment, and I do see people coming in now looking only for investments.

"To make a farm cash flow today you need to farm more acres. Prices haven't changed, but technology has increased yields and given us the ability to farm more acres. We have better crops and better chemicals. But with prices so low, the only thing a farmer can do is produce more bushels to potentially generate more income. You can talk with people about how they're doing, but you don't get an accurate picture until you start walking the fields, working out the bottom line, looking at input costs. That's what hurts the most, the input costs. If you're looking at getting bigger, you'll be running machinery across more acres and increasing your costs. But the trade-off is to provide the farm with more opportunity to produce.

"The scary part of farming is, when you look at the big picture, there's only one place a farmer has anywhere to move to stay ahead of poor prices and negotiate his own future, and that's land. There's always going to be a market for good land. If a nearby farmer sees a chance to get good land, he'll buy it or rent it, because land is hard to get. There are still enough good operators around to take over land coming available after another farmer has given up. Some farms may be split among three or four different farmers, so four farms get bigger by 500-acres rather than one farm getting bigger by 2,000-acres.

"The cycle of farming today is similar to that in the past — one generation establishes a farm, the second generation builds on it, and the third generation more often than not loses the farm or gives up farming. That's the trend. When the second generation retires, it is very likely the land is treated as a retirement investment so that the third generation will have to pay the going price for land to meet the retirement needs of mom and dad. Times have changed. Machinery and chemical prices are so high, the third generation can't farm the way dad or granddad did. Families still try to keep the farm in the family by cash renting. The farm is basically together — the ownership hasn't changed. The opportunity still exists for the family to farm it again, so they don't see renting as failing completely.

"There is a lack of sympathy out there among older farmers who during the '60s and '70s could farm virtually any way they wanted to and still make money. Prices were good and land values increased rapidly. After turning a farm over to the next generation, who are working harder and farming smarter and still not making any money, the older generation can't understand why they aren't making it.

"When people get desperate, they start grasping at things. They try exotic commodities like elk or emu. The prevailing alternate opinion is to spend more time managing the farm to be money ahead. Rather than looking for the quick fix, farmers should do things like soil sample to lower fertilizer costs, or know where to find an extra nickel per bushel. Tighter management will always provide farmers with a better opportunity for success. We have to keep in mind that farming is the only industry where the commodity sold has a price established by someone other than the producer. There will always be problems with farming simply because we are at the mercy of not only the weather, but political uncertainities as well." ∎

Jon Norstog was a county agent in Logan County before joining the farm and ranch asset management team at US Bank. He is responsible for 150 farms and 250,000-acres of crop and pastureland in North Dakota, Montana and Canada.

Sharold Geist
Durum Farmer ▪ Hazen

Sharold Geist has served on the North Dakota Wheat Commission and has seen the impacts of world grain marketing firsthand. Today he farms 3,000-acres with his wife, Dorothy, and prefers to be optimistic about agriculture.

"I raise durum to feed people. I sell all of my crop every year because I want to be a reliable supplier.

"While serving on the Wheat Commission I came to see what a big picture there is for wheat in the world. You realize you're working to give people their daily bread. I've seen the human end of the wheat chain and how important our grain is for the world. We only use half the wheat raised in the U.S., the rest goes to world markets, so marketing has to be an on-going effort if farmers are going to get a decent price. We either have to be a player in international markets or raise less wheat.

"We could be competitive in world markets if we used the tools we have available to us, like the Export Enhancement Program. The problem is, we start a program and then some political situation causes us to withdraw it. Other countries just step in and undersell us. When you travel around the world, people tell us American grain is too expensive and our farmers don't want to sell it anyway — we like to sit on our grain and wait for higher prices.

"Half of farming is luck. It's a matter of timing. You pick the right crop, get it in before it rains and sell it at the right time. A lot of guys farm through the rear view mirror. They see what worked last year and want to try it. But that was last year – it's probably not what's going to work this year. If you jump from crop to crop, year to year, you're always guessing and probably guessing wrong. I've decided to only raise durum and ride out the highs and lows.

"I've made changes as equipment and technology have improved. Today I'm totally no-till. Once I made the switch I wondered why I didn't do it sooner. My crops improved. I consistently raise more bushels year in and year out. No-till conserves whatever moisture we do get out here. I'm letting the crops get all the moisture that's available.

"When I bought a Concord air drill, most of the equipment I had instantly became obsolete, so I sold it off. I use a computer on my tractor to apply the right amount of anhydrous ammonia and put down seed and dry fertilizer in a one-pass operation. The first spring I used the Concord I came home and told Dorothy, 'This absolutely cannot work - it's too easy!' On land where I averaged thirty-five bushels an acre by conventional methods, I got a yield of forty-five bushels the first year with no-till — with less cost. I'm raising better crops today with no-till than I did conventionally. After one of my neighbors saw that, he also switched to no-till.

"I'm not afraid of new technology. I try to stay positive and optimistic about farming. I stop by the Extension Service, go to meetings and talk with other farmers. I soil sample every acre. My custom harvesters use a global positioning satellite (GPS) system and yield monitors so I know what each field is yielding. Every year I work with Farm Financial Services to determine my costs. We put all the inputs onto spread sheets and anticipate a price per bushel. I know my costs and projected profit scenario before I crawl on the tractor. I know where I'm at if it all works out. But this year I projected $4.50 durum and we're only getting three dollars so it didn't work out. I need some good years to carry over the bad price years like this one. Last year I had my whole crop sold by the second week in November. The bid was for $5.85 so I knew right away where my profit was and I sold.

"There's a lot of people to feed in the world. If our politics were right, if we could use the tools we already have to move our wheat and retain markets, if we could establish a quota system with Canada, if we could do the things that are at our disposal, I'd be very optimistic about farming. If the system we have were allowed to work, there would be food, markets and price for all. But politics and trade issues prevent the system from working.

"Farmers are caught in a bind. There's only so much you can do to control inputs. You try not to waste money, but there's a given cost per acre that you have to spend to put in your crop. It's frustrating when someone else sets the price. It's hard to give up a lot of retirement money with a three dollar crop. I lost about $80,000 this year. That's money that could have gone into a mutual fund. But there's not much you can do.

"I'm not sure how long I'll continue farming. My children don't want to come back and be risk takers. I have four neighbors who will be getting out in the next five years due to age, and there are no prospects in their families for taking over their farms, either. In the near future, who's going to live out here and do the work without it turning into a big central operation, something like what Russia had? What's the difference between government ownership of everything or a giant like Cargill owning all the farms? They won't farm for nothing; they want to make profits, so everything will be more expensive and less efficient if the giants take over farming. The United States has the most efficient farmers in the world. People should care that they survive." ▪

The stress is evident on the face of Sharold Geist as he waits through yet another late May rain before attempting to seed his fields. He already knows he will lose money on the crop he has yet to plant in the ground.

Rob and Jeri Dobrowski
Working Off ▪ Golva

Rob and Jeri Dobrowski are from farm and ranch families and seemed destined to farm themselves. They were both state FFA officers in Montana and were enrolled in ag programs at Montana State University when they met. They married in 1977 and moved to the farm near Golva that Rob's parents had operated. They began working off the farm to make ends meet before eventually deciding to give up farming. They participated in many programs designed to help farm families make the transition into other careers.

"By the spring of 1984 our dairy was paying for itself, but the farm operation as a whole wasn't. We were asking the cows to pay for our farming habit – buying our seed, fertilizer, and fuel and paying for everything. As a whole picture, we were going backwards. We agonized for a long time about what to do. We couldn't sleep. We'd sit up talking about it. We had just borrowed big bucks for equipment, to buy cows and put up a bulk tank room. We hadn't expanded much, but we still had borrowed a big chunk of change. Eventually, we decided to get out; it just wasn't working.

"We didn't have a banker or our lender tell us; we made the decision on our own. Our line of credit was getting higher and higher, and the payoff was getting lower and lower. Our net worth was declining every year. We were only in the dairy business for two and a half years, but we're still paying for it. If we had quit farming two years sooner, it would have saved us hundreds of thousands of dollars. Here we are today with twelve years still left on our note.

"When it comes down to it, the biggest thing about getting out is after you finally make the decision, everything gets a whole lot easier. Until then, the uncertainty of things is stressful. Once you make a decision, things get easier because then you have a direction to go in. You need to make a decision and move on.

"We talked about all the feelings of giving up the family farm, the farm that had been in the family for many years. We thought about it in business terms, but the emotion of the family farm is still there. Dad farmed this place since 1926, and he made it. The only way we could justify it was that things were different then for return on investment. Granted, they worked hard, no doubt about it. One year our uncle bought a piece of land, planted it fence line to fence line with barley, harvested it, sold the crop and completely paid off that land with one crop. You can't do that now.

"The biggest problem we had was that in our minds we had failed. And we can't tell you what we failed at specifically. Before we decided to quit, we figured if we worked harder and worked longer we could make it work. But it doesn't happen that way. Extra hours won't change anything if the price isn't there. It just won't. It's extremely difficult emotionally. Mom took our quitting better than dad. Time has helped us get over it. We did a Val Farmer workshop and attended mental health association workshops. We're glad there were places to go for help and moral support when we really needed it.

"We've always talked openly about our farming problems. It helped us to talk about leaving farming, but it bothered a lot of other people. We had people tell us we were crazy, that we were quitters. We had people tell us we hadn't tried hard enough. It's really hard when you have close friends and neighbors tell you you're crazy and ask why are you doing this? We imagined that in their minds they were thinking, `If it could happen to them, it can happen to us.' They were getting scared. But we also had a banker friend of ours tell us we were the bravest, smartest people he had ever come across. His partners couldn't believe there were people who had the smarts to get out while the getting was good.

"If you're raised a farmer or rancher, you think you're always going to be a farmer or a rancher. Having two or three careers in life is for someone else. These notions are true in our families and our community. We had been directed toward farming our whole lives. But we saw we had options. One friend told us he'd like to get out, but he's too old and can't do anything else. We hear that so often. People really sell themselves short.

"When we decided to get out, we sat down and made a list of everything we could possibly do. Nothing was beyond consideration. We had a long list. We started out driving truck and working for the local newspaper. (Rob now does finish carpentry and cabinet making. Jeri is a free-lance writer and works at Home on the Range part time, which helps pay medical insurance.) We fought to have flexible schedules so we could see our kids' school activities, their volleyball and football games. Many jobs force people to miss so much of their kids' lives.

"We've talked long and hard trying to discourage our children from farming. We're both from farm backgrounds and we wanted to raise our kids out here on the farm, but we wouldn't wish farming on either one of them.

"We're so happy we're out and we're not having to go through all of this turmoil in agriculture now. We're seeing friends and family members who are going through it. Sometimes we think some people are in denial of reality. They still haven't got it. They just can't keep on doing what they have been doing, but they haven't figured it out yet. It's frustrating. We feel impatience with them, a little anger and sympathy, because they need to take a hard look at their situation and make a decision, but they are avoiding it. They risk getting so far down they'll have absolutely nothing left. It's still difficult for us. It's not perfect. We think we may have to move to find better jobs soon, but things are so much better than what it was. We feel sorry for people who are having problems now, but we know they will make it if they try." ■

Rob and Jeri Dobrowski take a morning walk near their home west of Golva. Once they made the decision to leave farming, they were able to clearly focus on their options. Rob eventually became a finish carpenter and cabinet-maker, and Jeri uses her computer to produce free-lance writing for regional publications.

Frank and Deborah Popper stand on a little used country road, the virtual embodiment their of prediction of the Great Plains emptying out of people in rural areas. The Poppers' research shows North Dakota is the most at-risk state, and they have visited and spoken here more than in any other Great Plains state.

Frank and Deborah Popper
Buffalo Commons ▪ Highland Park, NJ

"From a meeting with Governor Sinner in July, 1988, our impression was he saw a religious dimension to the whole thing. When we think back, he thought there was something sacrilegious about what we were saying. When we were in his office, we were still at the point where we were absolutely amazed anybody had noticed our idea. Our article had made the rounds in state government and portions started appearing in his speeches. In truth, he did us an enormous favor by commenting on it — well, the word now would be demonizing, but I don't think anybody used that word back then. That sort of elevated us in ways we probably didn't deserve, and it just snowballed from there.

"In those days we were greeted with many negative questions and comments. 'Why are you out here? Are you here to drive my property values down? Are you here to do a land grab?' Now we can see that farmers, ranchers and energy interests saw Buffalo Commons as an assault on their livelihoods, legitimacy and even their ancestors. We heard negative reactions to our ideas much more frequently earlier on when people didn't know us.

"Buffalo Commons is really a question about how to talk about the future. If the trends of depopulation continue, what would the future look like? And so that's why we started to come up with something more evocative. Our writing was very much in images of what the Plains would look like. The images obviously couldn't be documented, but everybody could envision buffalo moving along the landscape. People who were having hard times thought the images were provocative. Environmental groups and Native Americans liked the image.

"Buffalo Commons would benefit the region by bringing in all kinds of buffalo production, buffalo tourism and buffalo culture, which would yield some major economic returns to the region that it's just beginning to get now. There's lots more coming. All this as cattle markets continue to decline and wheat markets aren't going anywhere. It offers something economical to these, in some cases, near death Plains towns in North Dakota and elsewhere that they just haven't had. How this will play out we can't really say. There are organizations that have sprung up in the last eleven years that didn't exist in 1987. They are kind of startling and interesting and inventive, and we suspect there are more of them on the way.

"In fairness, we can't point to a Plains community that has really gone full bore into Buffalo Commons — that is doing its planning and economic thinking around the Buffalo Commons. It doesn't exist yet. But we expect there will be such places emerging over the next decade. We think they are on their way because of the economic potential.

"We are a little waffly on whether Buffalo Commons will stem the population decline or not. We really don't know. We think one of the reasons why we get resented is that Buffalo Commons sounds to local ears like a way of accommodating further population decline without actually doing anything about it and just sort of accepting it. It's like, 'We're sliding down the slope and these Easterners have come up with a fancy name for it. It won't make any difference, we'll keep sliding anyway.' A tone like that. We can't point to any place that Buffalo Commons has saved. All we can point to are possibilities and emerging organizations that are experimenting with different alternatives.

"Usually we talk about the good Buffalo Commons. There's a possible bad Buffalo Commons, too. One is it gets taken over by agribusiness and twenty years from now we can go into McDonald's and buy a buffalo burger that's been processed by five or six companies that own the Great Plains. That would be a very plastic Buffalo Commons. Another one would be a sort of Disneyfied Buffalo Commons, where a lot of the Great Plains would be a sort of theme park. Another is that the Great Plains gradually becomes deserted and the federal government doesn't step in and the Great Plains basically goes to hell. None of these outcomes would be real happy ones for local people.

"Clearly, North Dakota reacted fast, and continues to react to our work. We're invited here more than any Plains state. Some of our projections show problems are worse in North Dakota or are of longest standing. North Dakota probably does have the most counties losing population and the largest actual area at risk. We think North Dakota was looking for ideas when the Buffalo Commons concept came along.

"Today we're a respectable minority, and the things we write aren't quite so provocative as the first things we did when we were probably a little more naíve. The reception to our ideas — both positive and negative — has made us more cautious, more responsible. Even without the drought of the late 1980's there would still be an argument for Buffalo Commons because of the outmigration of people, the aging population, poor farm prices and lack of investment here. It's all connected. It's just that after eleven years we're seeing less opposition or hostility to our ideas." ■

In December 1987, an article by the Poppers, "The Great Plains: From Dust to Dust," created a sensation in North Dakota. Its authors suggested the federal government buy farmland and convert it to pasture for grazing buffalo as an answer to the depopulation of the Great Plains. Frank Popper is a professor of urban studies at Rutgers University and Deborah Popper is a geographer at the College of Staten Island.

Kenny, Cindy, Michael and Matthew Gross (with Sparky and Pepper) pose by one of their six hundred-ton stacks of alfalfa hay from their second cutting. They maintain a website to help market their hay, and a typical 300-cow dairy will use at least one twenty-four ton truckload each month.

Cindy and Kenny Gross
Alfalfa Farmers ▪ Edinburg

"It's scary to go into something you don't know a whole lot about, and to invest a lot of money into something you haven't done before. We had been strictly small grain farmers, but we decided that if we were going to stay on the farm, we needed to do something different. We planted 500-acres the first year, which is a big step. Last year we planted 500 more acres. We farm 3,000-acres, so alfalfa is a third of our farm.

"Alfalfa is a whole lot more work than small grain farming. We take three cuttings off each summer. That's a lot of labor. During haying we have eight to ten people working in the field. Some are trainees – students who come over from countries like Finland or the Netherlands to work on farms and learn about farming practices in the United States. We also have one full-time man and another who works part time. And our boys help. Once hay is ready to be baled, we work at it really hard and fast to get it under a tarp or in a building so it doesn't take on any moisture.

"We sell most of our hay to dairy farms in Minnesota and Wisconsin, with some in Iowa, Illinois and Michigan. The dairy farmers will mix our hay with corn, their own silage, or cottonseed oil to make the proper ration they want to feed their cattle. For milk cows, farmers want hay with the highest relative feed value. It's a challenge to put up hay and keep it at a high relative feed value. We want to retain the leaves and have nice stems and not have mold or ferment from the field. We're looking for the right amount of protein, potassium and calcium. We have moisture testers on our balers and a spray bar that sprays a special acid onto the hay so that it won't mold or rot. It's kind of high tech, considering what the old balers were like.

"Raising alfalfa is not like grain farming, where you wait for the right price before hauling your crop to the elevator. We have to advertise in newspapers or at farm shows in other states. We have to contact the dairy people directly because there is no system for selling hay. My husband or our hired hand will actually go and meet potential customers. This way we have a good chance of setting up a good relationship – much better than if you just send hay out with a trucker. We've taken several bales along to Farm Progress Days in Menomonie, Wisconsin. We just sat on the bales talked with dairy farmers who came through the exhibits. Then we drove through the countryside, stopping at dairy farms – your basic door-to-door selling. We did that for about a week. We've found with hay, we spend most of the summer putting it up, then we're busy all winter marketing it.

"We're hoping to get steady, yearly customers so we won't have to be looking for new customers every year. We don't want to wonder whether or not they will pay on time, or if they won't accept the hay when it gets there. So far it's worked out okay, but you hear all kinds of horror stories with stop-pays on checks and checks that bounce. That's one reason why we like to sell our hay on contract. We have quite a few more contracts this year than last year. Some of our customers contract for a load a month, some twice a month. I think dairy farmers like to get hay from the same dependable source each time.

"Price was one of our reasons for getting into alfalfa, but lately prices have dropped. Last year there was a lot of hay, mostly because we got plenty of rain. Some producers in South Dakota and Minnesota took five cuttings off their fields – that's what causes prices to drop. Now, more Canadians have gone into hay production, and a lot of it is coming down here. We see trucks loaded with hay go by on the interstate, and it hurts us a lot. But you can't blame them – they are having a tough time and are trying to make a buck, too.

"Value-added agriculture is a possibility for us. We recently formed the Northland Alfalfa Producers Association, and we're looking at different types of processing for alfalfa. One way is to compress it so it can be cost effective to ship farther than we do now, maybe even to Pacific Rim countries. One association member traveled to France to visit a plant that extracts nutrients from alfalfa to make feed for livestock. The process starts by taking wet alfalfa right off the field without even baling it. That's the kind of processing we might have to do if alfalfa prices don't rise. The secret is to find a niche to fill with our product.

"In the last few years I've heard more people talking about not wanting their kids to farm. They don't want them to have to deal with the stress and uncertainty. But I think kids that are meant to be farmers will be farmers. The kids that have an attachment to the land and really enjoy farming will do it, if it's possible for them. That's probably a healthier way to look at it. Our kids will have to make that decision for themselves. We've decided we won't push farming on them." ▪

Cindy Gross grew up on the farm near Edinburg that now belongs to her and her husband. A graduate of Minot State University, she spent seventeen years as lab supervisor at the Park River hospital before deciding six years ago to spend more time on their farm. The Grosses have three children: Stephanie, and twin sons Mike and Matt. Four years ago, with grain prices staying low, they decided to grow alfalfa as one of their main cash crops.

Bill Patrie
Cooperative Developer ■ Mandan

Bill Patrie is the rural development director for the North Dakota Association of Rural Electric Cooperatives, and is the director of the Dakotas Cooperative Development Center. Patrie has been involved in the start up of many producer-owned, value-added agricultural cooperatives.

"Since 1991 there have been twenty-eight new generation cooperatives formed in North Dakota. These are twenty-eight different ideas and companies. Why have so many people been getting together in cooperatives? It's almost directly related to one word: hope. What else is there?

"Hope is a different way of thinking. Hope makes alliances possible. People can organize around opportunity rather than a perceived enemy. It energizes people and takes their fear of failure away. Hope allows people to join in and commit resources at their own speed. You can achieve a commitment without conformity, and you can celebrate diversity without trying to make everyone look exactly the same or believe in exactly the same thing.

"Anger, on the other hand, doesn't seem to work as a catalyst to bring people together. A negative vision only allows people to strike an emotional blow and get some instant relief. They feel like at least they've done something. But can you get a $20,000 investment for your own packing plant from anger? No. That only comes from a positive vision. People have to believe in it and really want it. The new producer co-ops that are working in North Dakota come from a disciplined business plan approach. They are successful because they were carefully researched and carefully thought out, and the investors had the discipline to stick with the whole program.

"I think the marketplace wants the kind of producer that our farmers and ranchers are – people who care about the end consumer. We are proud of the way we grow our crops and raise our animals. Consumers want to know we produce quality and that we care about our livestock. I think the people who buy our products are way, way more like us than we give them credit for being.

"I think you save rural America by trying to do something else – you look for opportunities that preserve our affinity for our state, our sense of nurturing and our love of people. If you get these three ingredients into your core business, you'll thrive. We need to find a vision that's so compelling people will follow it. A common vision in North Dakota almost always has these three elements. You save rural North Dakota by not trying to – because the harder you try to, the more surely you will fail. You focus on something else. You look for a way of thinking that generates cooperation and opportunities that preserve our sense of place and our love of people.

"We identify with places that are familiar to us. We respond to certain physical features, like cottonwood trees or a river we live by. Way more than we previously thought, our sense of place is terribly important to us. North Dakotans also have a love for nurturing and growing things. Especially in western North Dakota, we flat out love our animals. Why else would we respirate a newborn calf? Being part of creation has non-economic rewards for us. The third dimension is that we want to share all this with people. Living here, we have an unusual kinship or appreciation for our fellow human beings that is not common to the rest of the country. Maybe it's because we are sparsely populated. Maybe it's because our ancestors came from vastly different places. Whatever it is, we like people and we want them to like us. We tell people we don't make much money, and yes, it's remote here and it gets cold, but we live a good life in unspoiled, beautiful country.

"I think all this underpins why cooperation works better here than in other places. It goes to our sense of who we are. We have survived some pretty tough times. We actually enjoy each other, and we're used to working together to get things done. We've always helped our neighbors. Cooperative enterprises get to be another excuse for us to come together.

"We have to overcome two basic barriers if we are to achieve a new future with any kind of organization. The first barrier is an inappropriate sense of limits. People say things like, 'We know we can't do it. This hasn't been done here before. It only works in other places,' that kind of thinking. I say this is inappropriate thinking because if we work together, we know we can get things done. The second barrier is a sense of unworthiness: 'I don't deserve to be successful.' There is a feeling that we should be humble and just hunker down all the time. It's our fate to dig in the soil, and we should work hard and be content to die with our heads just a little bit above water.

"I think this movement toward cooperation is giving us hope. At least the dialogue has started and we're beginning to think differently. By working together we can preserve our core ideology, our way of life out here. North Dakotans have a love of place, a love of people and a love of nurturing. Cooperatives can bring these traits together and generate a solution for rural America and rural North Dakota." ■

Bill Patrie consults with farmers and ranchers anxious to form "new generation" cooperatives in the hope of diversifying their operations and obtaining some control over the price of their product.

Doc Throlson stands by Holly, one of his first buffalo heifers. The founder of America's only bison cooperative got into raising buffalo to relieve stress from his hectic veterinary practice and to raise an animal that was compatible with his lifestyle and farm property.

Dr. Ken Throlson
Bison Rancher ▪ New Rockford

"What's good about buffalo? How much time have you got? My gosh, there's so much that's good about buffalo. I definitely feel the meat is lower in cholesterol and has less harmful fats than other meat. It's good for you nutritionally. With beef, the flavor is in the fat; with buffalo, the flavor's in the meat. Buffalo will do well on poorer pasture or with poorer feed than cattle will, and they don't need sheds. Buffalo weather well, they will stop calving if a storm hits then resume when it's over. Beef cattle can't do that. Buffalo are just naturally suited to this environment.

"Buffalo are easy enough to handle once you figure out how they want to be handled. It's as simple as that. Once you know what they want to do, you won't have any problems. When I started treating them like an animal in the wild – letting them graze on natural prairie grass – it was like a light bulb went off in my head. They did better, looked healthier and gained weight faster. Buffalo don't need us as much as cattle do to survive. This animal can adapt.

"I gave seminars on buffalo ranching for years. Still do, in fact. My phone rings all the time. People come out and ask about getting into buffalo ranching. I tell 'em that I put my six kids through college just on what I earned from selling bones, skulls and hides. It's just been in the last few years that the bankers have started believing in buffalo. Now they're willing to loan to producers. There's a lot of people getting in the business. Everyone is trying to diversify in order to hang on in agriculture these days. In this township alone there are now seven bison ranchers.

"Another positive with buffalo is world trade. We all know markets are global today, and here's a chance for us to load the deck on our side. With buffalo, we have a commodity no one else has. Ninety-eight percent of the bison in the world are in North America, and eighty-five percent of those are in the United States.

"So little was known about buffalo when I was in vet school. Even today there's very little taught. I've been asked to help fill the void about caring for buffalo by speaking at continuing education classes and at vet schools. I'm not a good speaker or real comfortable with it, so it's hard for me to do, but there's so much demand out there. I've been raising buffalo longer than most people have, so I get called.

"Farmers don't need to get into buffalo in a big way in order to diversify their operations. I tell people to just take the acres they are wasting and run buffalo on them. Put twenty or thirty cows on some tough pasture land or land you're not using, and they'll pay for themselves. Anyway, smaller units of buffalo are easier to handle than a herd of 100 or 200 head.

"We've got to be careful about marketing these animals. Our cooperative has grown much faster than we anticipated. We've really had tremendous growth. But we have to go for quality. We can't just sell off our cull animals and expect people to like eating tough meat. Right now we're focusing on getting people to try buffalo in first-class restaurants. The meat is so lean it's less forgiving to cook than beef. A good chef will know what to do whereas a household cook might have trouble. I believe our industry will self-destruct if we don't take charge of it. It's in our best interests to train chefs and teach people how to prepare buffalo. And we have to consistently deliver a high quality product.

"The bison business is everything I believe in. It's natural. It makes sense. You know, the species that didn't adapt aren't here anymore. The buffalo is a survivor first and a producer second. I started out with two bulls and a heifer calf that cost $400 a piece. Now a heifer calf is averaging $2,600 while a beef heifer is bringing $300 or $400. That's a good return by any measurement." ▪

"Doc" Throlson is a pioneer in raising buffalo and was a leading advocate for organizing a producer-controlled industry that is today known as the North American Bison Cooperative. He is sought-out by producers worldwide for advice on raising buffalo.

Jim Harmon
Fourth Generation Farmer ■ Carrington

Jim Harmon is the fourth generation to operate a farm started by his great grandfather, who came to Dakota Territory in 1879. Harmon is past president of North Dakota Farm Bureau and has been involved in the leadership and policies of the organization for the past decade.

"My great-grandfather came to Dakota Territory as a surveyor for the railroad and became the first sheriff in this area. He started farming in 1883. I'm raising the fifth generation, hopefully, of farmers on this land. I want each of my three sons to have the opportunity to live the life I do.

"I'm still optimistic about farming. I'm confident we can change things and swing the pendulum back so we can be profitable and farming will be fun again. There's no better place to raise a family than on a farm. There's really no better work, either, no better way to make a living. It's kind of a miracle occupation because you're dealing with land that's just alive. Putting seed into the soil, watching them germinate and grow, nurturing a crop to do well, helping nature along but all the time being at the whim of nature, is exciting and gratifying. The trick is being able to stop and realize how beautiful this land is. Sometimes we get so hectic and concentrate so much on everything that is going wrong we forget to stop and look at everything that's going right. I've got a great family. The crops are beautiful, the sunflowers are blooming and wildlife is abundant. It's really like living in a utopia. Once you stop and think about that, it balances out some of the bad.

"The Aviko potato plant in Jamestown is here because we started a cooperative to get a plant in central North Dakota. The plant gave us the opportunity to have irrigated potatoes here. It really expanded our opportunities. It was a large capital outlay to expand into potatoes and in the beginning it took a lot of my equity to get into. It was a big risk, but the bigger risk was doing nothing. I was just breaking even – if that – by doing the same thing. There really haven't been any high profit years in the past five or six years, and I really didn't see any profitable years in the near future in conventional farming. Our 500-acres of irrigated potatoes is keeping us in the black.

"I think there are areas of hope in farming. Sometimes you manufacture your own hope. I've tried to do it with irrigated potatoes and joining the processing arena with Dakota Growers Pasta, which has been nicely profitable. There's been good opportunity if you can afford to invest. When times are tough in farming it's hard to find the capital to invest in value-added ventures. Either you get money the old-fashioned way and inherit it, or you try to borrow. It's real tough. The ones who did have enough capital are sitting pretty good when something like irrigated potatoes comes along and they can capitalize on it. It's kind of the case where the rich get richer. There are some programs where you can borrow, but not nearly enough. There have to be more opportunities for farmers to get capital so they can invest in value added.

"Another opportunity is to draw down our grain stocks through aggressive export programs. We've got to reclaim our worldwide market share. We don't seem to have the intestinal fortitude to take on the European Union. We absolutely must have an aggressive export enhancement program to fight them. We don't use what we've got now. It's all held back by government policy. The rest of the country is enjoying a very prosperous time. Part of the reason for this prosperity is a very depressed food market. Farmers are subsidizing cheap food in this country. If we're not going to use exports and let grain stocks rise, the price is going to continually drop for farmers dramatically. It's cause and effect. If you're going to make the effort – or lack of it – to let stocks rise and not export; government should be compensating farmers for subsidizing the rest of the economy.

"Farm Bureau is very disappointed in the lack of world market share for our grain. We went from storing ten percent of the world's wheat supply to where we now have twenty-five percent on hand. We're essentially becoming the world's grain bin again. It's been devastating. We're trying to correct that. We advocate an export deficiency payment. Basically, we're demanding that we have four billion dollars in export enhancement and that we use it. If our government, for whatever reason, decides not to recapture our market share, then we advocate a scaled program where for every ten million bushels above 360 million bushels on hand there's a direct payment to the farmer of three-point-two cents per bushel. If we're not going to export, if we're going to let our stocks rise, if we're going to let our marketing process fail, then farmers should be compensated for subsidizing the rest of the economy.

"I don't think you can lay all the problems farmers have on one cause. I think it's a combination of all the things we face today – low prices, high input costs and high prices for machinery. But I think there are some villains here and they are the ones who make our trade policy. When these people go in to negotiate a GATT agreement, for instance, the farm business is not the highest on their list of priorities. In my opinion farmers are sacrificed for other entities. That has to change. Agriculture has to be at the forefront. We should at least be treated on equal footing. It's a mindset, I think, with our government that agriculture be put on the back burner on trade issues.

"I think farmers get a voice in these matters by being louder and being more active. We've got to shake a few trees. When we come into hard times like this, sometimes its more tempting to drawback and not be active. I think its essential that everyone get more active and let everyone know how we feel about this situation. We need to let the rest of the country know why it's in their best interest to keep farmers in business and keep us profitable." ■

Jim Harmon would like to pass on the family farm to his sons - Brett, Ian and Kellin. Harmon's wife, Maria, operated a bed and breakfast in their historic home for several years before taking a job at the nearby Dakota Growers Pasta plant.

Dan Wipf sits in his office where he uses a powerful computer to design custom roof and floor trusses.
Until this unusual business opportunity came along, the Willowbank Hutterite Colony was having difficulty making
ends meet solely with their farming operation.

Dan Wipf
Willowbank Hutterite Colony ▪ Edgeley

"There's sixty communities like this one in North Dakota, South Dakota and Minnesota, and we all operate about the same way. When we started this place in 1978, we had 2,000-acres of land, but interest rates were really high, exorbitant even, at eighteen percent. We just couldn't make it by farming, so we gave our land back to the federal land bank and got out of farming. But we had to do something for a living. We continued with our dairy cows and raising turkeys, and by some quirk of fate, some accident, we stumbled onto a truss manufacturing plant. So we started on the bottom rung of the ladder, building roof and floor trusses and worked our way up to where we are now. Eventually we did go back into farming, basically because we needed a place to put our manure from the turkeys and dairy.

"We've diversified. We have enterprises. Every enterprise has its designated managers who oversee the workers and keep track of expenses and revenues. We have a farm enterprise, a turkey enterprise, a chicken enterprise and the dairy enterprise. We raise mostly corn and soybeans on the farm. With the corn, what we don't feed we sell. We milk 100 cows twice a day. Our most profitable enterprise and our main source of income is our truss plant. We sell trusses to large contractors and dealers all the way from Colorado to Iowa, Minnesota and Montana. All custom orders. If it wasn't for the truss plant, we couldn't have bought the land we own now.

"We raise 100,000 turkeys a year and sell them to a processing plant at Watertown, South Dakota. Three guys take care of the turkeys and another three guys do our milking. The trusses are labor-intensive and so most of our people work on trusses. We have ninety-two people living here, infants to adults. Everyone has a job — is given a job by majority vote. It's a system that's been working for our order since 1528.

"Basically, the New Testament sets our guidelines. Our communal way of life is based on the second chapter of Acts, where it says all those who believe sell your possessions and lay your money at the apostles' feet, and everybody will be given what they need. This is how we base our way of life. Outside of that, we try to live by the whole of the New Testament. No one in this community has any money or owns anything.

"As far as our entity goes, I am the president and my spiritual status is minister. We have a secretary-treasurer and board of directors just like any organization. But if I want to spend money in excess of a certain dollar amount it has to be okayed by the group. All the married men vote. If we need a new pickup truck for everyone to use, we vote on it. I'm elected for life or as long as health permits and I abide by the rules of the good book. It's just turned thirty years that I have been president. We believe that belief comes before baptism. That's

our philosophy and the teaching we go by. Our education – what we learn in the community and in the home – is scripturally based. Few people from the outside ever join us. Years ago in the 1950s we had some people join the colony. But I guess it's pretty tough to completely submit yourself and say no to yourself and give everything away. I suppose it's hard for someone to accept our way if they weren't born into it.

"If we were to only farm it would be tough to make a living. I'm sure we couldn't do it because we're trying to feed ninety-two mouths on the land we own. I don't see how it's possible for anybody who lives any higher than we do to make a living on farming alone. In my opinion, if a farmer around here tried to make a living on 1,000-acres he would be fighting a losing battle, especially if he had to borrow money to pay for debt. We have some advantages in numbers. There are averages that are in our favor. Our situation is labor intensive, but our work is pooled, so if one enterprise loses money, maybe another one will make a little. We can probably get by with less than some farmers do because of our simpler way of life. If we were just farming, trying to make ends meet on just the land, we'd probably need 10,000-acres to make it.

"We see it's true that farmers have to diversify. I don't feel that we have the answer, but I do believe if a farmer can't diversify, he's in a pretty volatile situation. If he's only into one thing, I don't know how long he can make a living. You need something to fall back on. We own 2,600-acres now, and we farm 4,000-acres. We could conveniently farm 6,000-acres, but we're not looking for more land. Our farming habit is big enough. We're geared to an operation this size with the equipment we have and the number of people we have to work the farm enterprise. That's where we want to leave it. We've bought quite a bit of equipment the last two years, and I know from a fact that if we had to purchase equipment from farm income alone, it would have been impossible.

"I always feel that a person, I don't care what his way of life is and what his occupation is, has to put something aside for a rainy day. A farmer, if he had a real good crop like we did in 1998, can't just go and splurge on anything. You have to be careful and reign in your wants and go by your needs. I'm sure most people left in farming are smart enough to realize this. In farming, the good years are the ones to look out for; you can overextend pretty quickly. We know of other communities that are big into farming and smaller in other enterprises, and I have to say they aren't in the best of straits. They're struggling because their enterprises aren't doing well. In farming, no matter how it's done, you need to diversify to ride out the ups and downs." ▪

Dan Wipf grew up near Mitchell, South Dakota, and moved to the Fairview Colony near Edgeley in 1970. Eight years later he helped establish this colony ten miles further west. The families farm 4,000-acres, maintain a dairy and raise turkeys and chickens. In spite of their communal lifestyle where everyone shares resources, Wipf believes they couldn't make it on farming alone. The inspiration for their Anabaptist order comes from Acts II:41-47.

Clarence Johnsrud
Retired Farmer ■ Williston

Clarence Johnsrud returned to Williston at age thirty following service in World War II. His first year farming produced a twenty-bushel wheat crop and netted him $1,400. Eventually he bought land in the Buford-Trenton Bottoms near the confluence of the Yellowstone and Missouri rivers where he began a successful farming operation, including irrigated sugar beets.

"Sometimes I think farmers are better off if they don't get high prices for their grain because too many others try to jump into farming. Look at the amount of land that's been opened up in South America. Those countries down there are just flooding the market with grain. There's so much luck involved in farming – twenty years without a hailstorm is luck. You need good weather and good bankers to make it, but you also need to be a good manager and have good workers.

"I'm real scared about the future of farming. It's a good thing I'm eighty-three, I probably wouldn't farm again – especially wheat – that's for sure. I've been very successful and I don't understand why. I've been able to hire good men; you've got to have good men. So much money is spent on inputs, and fuel and machinery, it takes an above average crop every year to justify that kind of spending.

"I think I'm a good farmer. I'm about as thrifty as you can find. When I was farming I didn't spend a dime if I didn't have to. I took real good care of my machinery. I only bought what I needed when I could afford it. If you asked me to start over again farming the hills up north here where I grew up, boy, I don't know. The land costs too much, the rent is too much or else the price of wheat has to come up in order to make it. We are so limited up here with moisture. This is dry land country, but it can be done with some breaks and some luck.

"On dry land today you may pay thirty dollars an acre cash rent. That's on every acre, so you can't afford to summer fallow. Then you'll pay fifteen dollars an acre to have it combined, and the farmer is left with nothing on his wheat after he's paid his input costs. It's hopeless. The price of land has got to drop to help the farmer. But if land drops then your collateral at the bank is shot to pieces. So the banks make it harder to get capital. Without someone backing you, it's a cycle you can't get out of.

"At what point do you give up? When you have some equity left or after you lose everything? From 1946 and for the next thirty years farming was real good. These are bad years now. It's real tough. I turned over my farm twenty-four years ago to my hired man, and now I work for him if I wish. Our son is involved with it now, too, and both of them have won awards from Holly Sugar for twent- ton yields with over twenty percent sugar. I still drive machinery, but I don't have the pressure of day-to-day farming. Anyway, us older guys need to get out of the way and let the younger guys take over. Our son and our renter have been successful because they had the opportunity. You've gotta give people a chance to succeed. The way things are in farming today, not many people have a chance to be successful.

"Farmers need capital to farm. The bankers need to support farmers so they can be in a position to succeed. You can't go to the bank and only get sixty percent of the operating money you need for the year. You are through if that happens. It costs hundreds of thousands of dollars to farm. You need the support of the lenders in addition to your own talent to make it. Can you see how tough it is for these guys?

"Here's what happens. Right now farmers have been hauling in last year's crop to the elevator because the loans have been called. Okay, so they've got to sell and the elevator is just jammed. The elevators are dumping wheat on the ground because they are so full. The wheat is only eleven percent protein, so nobody wants it. The elevator will pay $1.60 a bushel, take it or leave it, plus a five-cent discount for loss from the pile outside. At that price it doesn't take a computer to figure out how hopeless it is for making any money on wheat. What do you do? Put up more bins to hold wheat until there's a better price? Sometime you gotta sell it.

"If the farming economy is so bad when the good operators can't make it, who is going to take over when us old guys quit? I'm not saying all farmers have to survive, especially those who aren't frugal or good operators. But something has got to give. You have to understand I started with nothing, and now our son and I own 700-acres of irrigated land without debt. That land has high ground water from silt and water backing up from Lake Sakakawea, so the federal government is finally buying us all out. They should have done it back in the 1950s when the problem was discovered. All the houses have flooded basements. They are offering us more than $2,000 an acre and we can farm the land until it gets too bad.

"We never depended on wheat. We raise sugar beets for our main income. Holly Sugar has told us they will close their doors if they can't make a profit. So we have to raise good beets with at least seventeen and one-half percent sugar or face the loss of beet acres. Without beets, the value of our land drops. Sometimes I think the cost of irrigated land was too high. But I took the financial risk to buy that land because I always believed the government would have to buy us out someday. For me, there is great deal of satisfaction in seeing a beautiful field of sugar beets without a weed in it, knowing you had some small part in making that happen." ■

Clarence Johnsrud has seen the good times and bad times in farming, and he's glad he had irrigated sugarbeet ground to farm. He spends considerable time today finding fossils in the bottomlands of the confluence of Missouri and Yellowstone rivers.

Bonnie and Brent Woodworth wear protective clothing and use smoke to calm bees so they can handle hives in the field.
Bees are vital for pollinating crops, especially higher value added crops like apples and nuts. Unrestricted competition from China and Argentina has severely impacted American beekeepers.

Bonnie Woodworth
Beekeeper ■ Halliday

"Working with bees isn't as risky as you might think. If it's spring and there are flowers blooming, a colony of bees will be in a good mood because they are foraging in flowers for nectar to make honey. You can work with bees in shirtsleeves or even shorts and maybe even without your veil if you are very careful.

"But on a dark, cloudy day, you will want to wear a veil, coveralls and gloves because the bees will not be out flying and likely not in a good mood. Fast-moving objects and dark colors irritate bees, so we keep that in mind as we work with them. How defensive bees are depends on the breed you're working with. Some bees are more likely to sting you than others.

"The pollination services honeybees provide are very important to agriculture and the production of food in this country. It's been estimated that the annual value of increased agricultural production attributable to honeybee pollination is more than fourteen billion dollars. Our bees spend more time away in other states pollinating crops than they spend here in North Dakota producing honey. Flowers quit producing nectar in the fall after frost, so then we truck our bees to California, where they are kept in the foothills of Yosemite National Park through the winter, waiting for the almond pollination to begin. Before the bloom begins we move the bees into the almond orchards in the San Joaquin Valley. Since honeybees are so necessary for successful almond pollination, growers pay us to supply bees for their orchards. Last year nearly one million colonies of bees were contracted for almond pollination in California.

"After the almond bloom is finished, we go out to California and spend up to six weeks splitting our colonies and getting them ready for their next job, apple pollination in Washington state. When our bees arrive there, someone we contract with delivers the bees to the apple orchards. After they have finished pollinating the trees, usually in May, our brokers return our bees to us in North Dakota.

"It's generally late July when we start collecting honey from the hives. The honey flow can last until September, and we will do the honey extracting – removing honey from the combs – into October. The honey is stored in fifty-five gallon barrels. We ship our honey to a packer in Ohio, where it is blended with honey from other producers and marketed on the East Coast.

"Many beekeepers develop an attachment for their bees, just as you would for a pet or for livestock you care for. We need to carefully manage the hives to prevent swarming, which happens in spring when the hives get too full of bees. We work to keep our bees healthy and free of disease and pests. Finding summer locations with lots of clover, alfalfa and other nectar sources for the bees is crucial for making a good honey crop, the ultimate goal of beekeeping in North Dakota.

"Beekeepers received a support price until a few years ago. With the last farm bill, price supports for honey was dropped. It was wasteful government spending, some lawmakers said. So beekeepers are now at the mercy of world trade. China and Argentina have become our main competitors in the world honey market, which has negatively affected prices and caused many U.S. beekeepers to go out of business. This will ultimately affect U.S. agriculture because we can always import honey, but we can't import the pollination work our bees provide.

"We can't compete with someone in Argentina who makes $300 a month, unless we want to live the way they do. I suppose we can live poorly in order to keep doing what we love, as long as we don't expect to live as well as our friends do who have quit farming and moved to the city. I don't see many young people these days looking for hard work and low-paying jobs. Most Americans aren't willing to go back to the 1950s and live the way people did then.

"From our experience with honey, I can see our food production shifting to foreign countries. For one thing, their people are willing to live on less than we do. But we have taught them so well how to produce commodities and compete with us – we've given them the tools, technology, everything they need to feed themselves so that they don't need us to feed them. And consumers in the U.S. are less likely to care where their food comes from, whether they're eating grapes from Chile or grapes from California.

"I see so many of our youngest, brightest farmers moving off the farm and into Bismarck or Fargo. I hate to see it. Yeah, they are doing better with a job in town, but it's just not fair, because all they really want to do is farm. But on the farm they can't provide for their families, not the way they can with a job in town.

"In this part of the country agriculture is always a struggle. It's really interesting that in European agriculture the farmers are the wealthy people. They are respected and protected. Here it's just the opposite. We're reducing our farmers to mere paupers." ■

Most years, North Dakota is one of the top three honey producers. The Woodworth family has a tradition of keeping bees, starting in Iowa, then continuing when they moved here in 1955. Bonnie Woodworth has served as president of the North Dakota Beekeepers Association and chair of the American Farm Bureau Honey Bee Advisory Committee.

Joe Satrom
Ducks Unlimited/The Nature Conservancy ▪ Bismarck

In March 1999,
following a
reorganization of field
representatives, Joe
Satrom left the Nature
Conservancy after
serving for more than
seven years as director
of the North and South
Dakota programs.
Satrom joined Ducks
Unlimited where he
continues to work for
land preservation
and conservation.

"One of the most successful policies at the Conservancy is whenever a Davis Ranch or Cross Ranch is acquired, we work to financially endow the property to protect the land in perpetuity. In other words, there may always be a source of money to manage the land. But institutions and organizations change and disappear. Frankly, I wonder if we can protect anything for perpetuity, given how much and how fast things change.

"My family has farmed on the edge of the Red River Valley on some of the same land for over a hundred years. Given the volatility of agriculture, I have the feeling agriculture as I know it, and as my father and grandfather knew it, will not exist as long as it did for the Mandan Indians. They farmed along the Missouri River for some 700 years. We won't make it that long because modern agriculture is just so much more volatile, our use of land is much more intensive, and crop production practices continue to change very dramatically.

"In the future, I think land ownership patterns will change because of all of the other changes in agriculture. Conservation is such a preeminent purpose and is so valuable in terms of long-term land use. Conservation organizations will need to become equity partners with some ranchers and farmers that own critical natural habitats. The Conservancy is a large organization —last year members contributed $257 million. So I can see the Conservancy or Ducks Unlimited for that matter, becoming long-term partners with people who own biologically significant land. We essentially will have to join in an equity position and bear the risk of helping the farmer or rancher figure out not only how to conserve biodiversity, but make it work in terms of sustainable economics.

"I think we have to figure out how we can raise sophisticated crops and still protect the biological diversity of natural areas right in the same square mile or within the same watershed or landscape. Currently we sometimes seem to be at loggerheads, but I think the farmer and rancher has more at stake protecting biodiversity than almost anybody because their livelihood is based on the land and natural systems.

"We turn down a lot of land that is offered to us. Usually it doesn't have major biodiversity values. Land needs biodiversity of significance in order for conservation groups to be interested in it. We pretty much know the land we'd be interested in. For instance, the Conservancy is looking at land in the Sheyenne River delta that has ninety-five percent of the remaining native tall grass prairie in North Dakota. Of course, as a total, that's only a few tenths of one percent of what was once tall grass prairie. The hope is to set this land aside as a nature preserve.

"The Davis Ranch in Sheridan County is probably the Conservancy's most spectacular site. It's 7,000-acres. Scientists would argue that we are on the minimal side of what is sustainable with that piece of prairie. There's a real debate in science about what is a sustainable ecosystem. I understand that a certain species of prairie songbird needs about thirty acres for a breeding pair, about as much territory as a cow and calf in some areas of the state. So we are moving toward preserving larger landscapes where we think we will have a reasonable chance of providing long-term protection of the diversity. With changes in global weather and without a lot of what we call keystone species, such as bison, prairie dogs, ferrets and so on, it's doubtful that a lot of species will survive on smaller tracts of land, because everything in nature is linked.

"Some tracts of native land are threatened. Cultivation is just a matter of time. It's an important issue. A lot of land conversion is driven by very short-term opportunity. An increase in grain or livestock prices causes more land that probably shouldn't be tilled to come into production. You can't dispute the right of the landowner to make the decision to convert grassland or the perceived need to use the land for cultivation. Our role, of course, is to preserve. If we expect landowners to forgo some of their economic opportunities in favor of conservation, I think we'll need to be involved in buying or figure out some other way to reward them for making a decision for conservation. This is being done in other parts of the country and it's working. Our main work now isn't acquisition. Ninety percent of our work involves conservation easements on grasslands or wetlands. We are buying restrictive easements on grassland so it can be preserved while still bringing a return of some kind to the landowner. In North Dakota we're hoping to get some relief from current law that restricts the use of conservation easements as a land protection tool.

"I think many people want to leave a conservation heritage. That's evident by the amount of land offered to us. People just very much want some tracts of land to remain as God created them. People want us to help them make sure the land will stay natural. Despite some limitations put on us by legislation, we've never been concerned that the public doesn't support us in forest, grassland or wetland preservation. People like to enjoy natural landscapes. The conservation community is changing to fit the reality of rural areas and rural life today. We've been modestly successful in protecting biodiversity, but to be more successful we need to have a much greater impact on private lands. Most of the private land decisions are made by people on the land, so in the future I expect groups like Ducks Unlimited and the Nature Conservancy to work more aggressively at the community level." ▪

Joe Satrom worked for several years with the Nature Conservancy to locate and preserve environmentally unique land, such as the Davis Ranch in Sheridan County. As a result, he gained an understanding and appreciation for the sensitivity of natural areas. He advocates protecting tracts of biodiverse land while still enabling producers to intensively farm their most productive land.

Doris Goettle
Conservation Ranching ■ Donnybrook

Doris and Jim Goettle
are in the planning
stages of an agreement
with conservation
organizations to
maintain a large part of
their ranch in its natural
state. Doris has been a
nurse in hospitals and
nursing homes at
Stanley, Kenmare and
Parshall. On their ranch
near Donnybrook they
raise cattle, sheep and
small grains. They have
five children, three of
whom are grown.

"Money isn't everything. If you destroy your land to make the almighty dollar, in the long run you haven't gained anything. You've only hurt yourself. We need to think about what's down the road ten years, twenty years and beyond. We continually try to look at the big picture and do what we can to take care of what God has put here for us.

"Land is one of the few renewable resources that we have in North Dakota. It will continually give back to us as long as we treat it the right way - don't overgraze it or continually crop the same piece of land again and again, or pollute the water system. Jim and I were both raised on farms, and we have learned to appreciate the land and be good stewards, no matter what we do. We use a rotation system with our cattle and sheep so grass isn't grazed down too far. If you overgraze you can end up with blue sage or buck brush in your pasture, which isn't good for much of anything. Because of our system, we've been fortunate to have good pasture even in dry years. We avoid using chemicals that would get into the water system. That's especially important out here because this is pothole country. Whatever gets into the water is going to stay there until the potholes dry up.

"We're working with the U.S. Fish and Wildlife Service on a ninety-nine year easement in which we agree not to break up this ground or use it for purposes other than what it was meant for, which is grazing. It's about twenty-five hundred acres. It's mostly pasture because of the terrain and rocks. We can still graze cattle and hunt the land for deer or waterfowl. Our easement basically says we won't do anything to disrupt the natural prairie. We don't think that will make any difference for our operation. Besides farming, this land could bring in all sorts of people – not only bird watchers but also hikers and hunters. People from cities want to come out to places like this to see and experience natural habitat.

"We're looking at alternative measures to use our land and earn needed money from it in a good way. Most of our land is virgin soil. A lot of it is pasture and we have some unique species of birds here. There's piping plovers out near the salt lakes and east of here are stomping grounds for grouse. We see bald eagles and golden eagles. There are things out here you can't even explain to people from cities who don't know the smell of spring or fresh-turned soil. We live with nature out here. So we would like to work with groups like Nature Conservancy or the Audubon Society. If people come out here and take the time to stop, look and listen to nature, there is a lot for them to see and experience.

"Today, no matter how hard you work, the money isn't there to buy even necessary things. There are machinery repairs and vet expenses. We can't do anything about these expenses; on a farm they're just there and you have to pay them – even when you are getting one fourth of its worth for what you produce. The farmer is the biggest gambler around and there is always a need for extra income. We're out here working twelve to sixteen hours a day and sometimes more. When you figure the number of hours a farmer puts in and what we're getting back, most farmers end up working for seventy-cents to a dollar an hour. When you see the minimum wage going up and farm income going down, you know something is wrong somewhere.

"If I took your paycheck and cut it in half, and said 'This is all you're getting this week,' that would hurt, wouldn't it? But that's what's happening to us. We're good managers and good stewards, but our paycheck has been cut in half and sometimes more. The prices just aren't there. We deserve better prices because we know our products are good. People are buying what we produce in grocery stores! The people producing food should be able to make a good living like anyone else.

"Take wool, for example. We should be getting a dollar to a dollar and a half per pound for wool. For the past two years it has cost us about three dollars a head to shear. But wool is worth so little, sometimes only ten to fifteen cents a pound. Yet we have the best and cleanest wool in the area – shearers who work all through the Midwest have told us that. It's really sad after you go to all the work to keep the wool clean, use the best shearers, bag it correctly, keep your animals tick free and then you can't get anything for your work and what you've produced. We also see it in lamb prices. When you only get forty to fifty cents a pound for lamb instead of ninety cents or a dollar, you just can't make it. But grocery stores are selling lamb for four or five dollars a pound.

"Our children know the value of living on the land and the kind of life we lead. They are the fourth generation of Goettles on this land, and we would like to see the family continue on this farm. We've been very open with our children about farming – our supper table is more like a boardroom where we discuss what's happening with the farm and what we're going to be doing the next day. Our kids have as much say as they want, and we take them seriously. Many times during calving and lambing they are the ones out there working. Not only can they run every piece of machinery, they can repair the machinery and they're good at it. If they have an idea for improving the way we do things, we need that input. There are fewer people out here to help than there used to be. Our family is used to hard work. We understand time management." ■

Doris and Jim Goettle walk on some of the miles of pastureland they own. They hope bird watchers and people interested in natural prairie ecosystems will come to their ranch to enjoy the experience. If so, it may add a source of income without adversely impacting their ranching business.

Richard Haugeberg stands against the dramatic backdrop of a blooming canola field, part of a successful crop rotation system he uses in his no-till operation. Haugeberg believes farmers' best hope for economic survival is diversifying into value-added ventures such as Dakota Growers Pasta.

Richard Haugeberg
Grain Farmer ▪ Max

"These days a farmer has to look at farming as a business. Every once in a while we have to step back and ask, 'How am I doing?' Then we need to ask ourselves, 'Is my business growing the way I want it to?' If it isn't, the next question we need to ask is, 'What can I do to change?' and then do something about it.

"Farming hasn't been considered a business by many people who are farming. In some cases it's been a family that has been able to make a living by doing what their father and grandfather did for years, planting the same crops and doing things the same way. For many years that may have worked, but not in today's farming environment.

"If farmers find that one crop isn't paying the way it ought to, then we should think about other crops that might be more profitable. In order to make sensible decisions, we have to keep close track of costs and profits and then study those figures carefully. Or if we are thinking about renting more land, we need to figure out things beforehand to see if we would really be making more money by planting that extra land. Planting more land doesn't necessarily mean more profits. Rather than doing the same thing over and over again, I believe we have to look at farming from the standpoint of what can we do differently. And we have to remember the bottom line. We have to make enough money to pay our bills and have some left over for our families.

"It took me awhile to decide if no-till would be right for my farm. It requires a different type of equipment, and you have to learn about chemicals for controlling weeds. I read quite a lot about it and decided it would be a good idea. For seven years out of ten, we're looking for moisture in this part of the state. My land is quite hilly. I like the fact that no-till helps keep moisture in the soil before seeding a crop. Although the majority of farmers in this area have reduced the amount of tillage they do on their fields from ten years ago, I doubt that ten percent of us are using no-till. It's not easy changing your farming practices.

"It's no secret that the farther up the food chain you go, the more profit there is for what we produce. The problem with simply hauling our grain to the elevator and collecting our check is that the lion's share of the money made on that grain will not go to farmers, but to the processors and marketers who take over that grain after we deliver it to them. That's why wheat farmers get so little from the grain in a loaf of bread on the supermarket shelf. The problem is we ship too many of our commodities off to distant places for processing, so most of the money made on what we produce goes to someone else outside the state.

"One answer is value-added agriculture, which is another way to treat farming as a business. I've been involved with Dakota Growers Cooperative at Carrington. Our durum comes off a truck or rail car at the plant and comes out the other end as a box of pasta that can go all over the country and around the world. So instead of letting the profits from processing and marketing our durum go others in, say, Minneapolis and St. Paul, that money stays right here in North Dakota, simply because we do the processing and marketing ourselves.

"Dakota Growers is one of the top value-added ventures in the United States. It's really been successful for farmers in this area. The plant has been milling 40,000 bushels a day. That's more than forty semi loads! Most of the durum is bought through a pool system set up by Dakota Growers with the elevators. They put out bids to buy durum at the best price and to meet their standards for quality. Dakota Growers members are also stockholders, so we profit not only from the durum we raise and sell, but also from the processing and selling of pasta. Owning stock in Dakota Growers is a good way for farmers to diversify their investments. It's what more farmers should consider doing.

"To deal with today's low commodity prices we need to really pursue exports. Lately, the United States, Canada and Europe have had bumper crops. But when our government slaps a political sanction on other countries, we lose out. Canada is a willing seller to anyone. Cuba, for instance. Cuba would buy a lot of durum from us if they could, but instead they have to buy it from Canada. And that Canadian durum probably comes right through here on its way to Cuba.

"If we're serious about making farming a business, we'll need to do more than deliver our grain to the local elevator and collect a check. We'll need to think more about who our customers are and how they will use what we are trying to sell them. Marketing is an area we're going to have to sharpen up on if we're going to be successful in this business of farming." ▪

Richard Haugeberg grew up on the 4,000-acre farm he now operates. He holds a degree in agricultural engineering from North Dakota State University. He is a member of Dakota Growers Pasta Cooperative, a past president of the U.S. Durum Growers Association and a member of the board of directors of the National Pasta Association.

Major General Keith Bjerke
Life After Farming ■ Bismarck

"Being selected as an outstanding young farmer was the fulfillment of a dream for me. Back then we had just come through the glory days of the early 1970s. Things were awfully good on the farm. Our net worth was growing, equipment was still relatively inexpensive and our returns were very good. It was probably the best period of farming I've ever witnessed.

"My dad operated the farm before me and brought it through the Depression. He was conservative and cautious. I was the epitome of the expansion crowd that came out of college with grandiose plans and ideas. We waded in with a lot of debt and expanded the farm ten-fold. It was working well until along about 1979 and 1980, when things started to tighten up. In 1989 when we purchased our last real estate we hit probably the all-time high for overspending on land. That land has since been sold at a loss because we had to clear some paper. We couldn't continue with the debt load we had. It took me until the mid-1980s to embrace the fact that farming is a business like everything else. It's a great and glorious life and you love to have your kids there and be a part of it, but I can tell you that equity melts away very quickly unless you put some business sense into it.

"We were slow to react. We depreciated a tremendous amount of equity. We stayed in too long and tried to ride out high-priced equipment, high-priced land and high input costs at less than break-even returns. We got out of it without going through bankruptcy but we're paying off debt to this day. Instead of saving up for personal retirement, we're paying back farm debt. For the first time now in thirty-five years, we are starting to see a light at the end of the tunnel. People our age are looking forward to Phoenix and fun, and we're not prepared for that.

"But I have been probably the most fortunate person ever to leave the farm in North Dakota. I had a fantastic job in Washington, D.C. for four years, and immediately after that I was invited by the governor to serve in this job. It was something I had never ever dreamt about or thought about. I thank God and a lot of lucky stars for bringing me back full circle.

"I've been a member of the military since 1958, right out of high school. I love the structure, I love the people and I love the flag and all the things that it stands for. But on top of that, it was a great part-time job. During those years of farm struggle my Guard check put a lot of groceries on the table. I wouldn't say it was a matter of necessity, but without the Guard it would have been a much tougher time for me.

"I often hear from sidewalk forecasters that any time now the big conglomerates will come in and farm North Dakota. Why? When they can buy grain at below the cost of production, why would they want to produce it? Farmers are so efficient that those few who are left in the business are able to supply the world at below break-even prices. Any corporation that has to live by a balance sheet would say, 'we don't want that part of the business.' People thought our wet conditions this year would have an impact on world prices. It didn't affect price one bit. The fact really is, whether North Dakota has a crop or not, it probably doesn't tip the scales more than a nickel or two. This market is so big and so global that someone, somewhere, is producing the product, and the conglomerates can get their grain anywhere they want.

"We coined a phrase when I was candidate for agriculture commissioner along the lines that farmers need to produce a product for a market, instead of expecting markets for every product. We have been very slow to respond to market niches. I think producer-owned cooperatives are fabulous if the end result is a marketable product. One of the most successful achievements ever in North Dakota is the buffalo business. Those producers are doing very well and there are two reasons for that. One is the very low cost of production, and second, it's a quality product that the world wants to buy.

"The most difficult thing for people to relate to in farming is the horrendous cost of land, equipment and inputs — the fertilizer, chemicals and seed. There's got to be a tremendous cash flow to generate payments for these high costs. In my early farming years the only mistake you made was not buying equipment and land fast enough. Everything was more expensive a year later and everything you owned escalated in value. Things were just booming. In fact, until the bubble burst, inflation was one of the best things to happen to agriculture. In 1979 when we had a heavy debt load on the farm, we paid twenty percent interest for money. We had a good crop and went to the bank and gave them everything we earned and it didn't even pay off the interest. So you add one year's debt to another at twenty percent and pretty soon things are out of control. Those years of high interest were just devastating on the farm.

"I'm still optimistic about agriculture. I know farmers who last year had the best year they've ever had. The problem is what works great for one person doesn't necessarily work for another person living right down the road. We are looking for answers in places where it's still an individual art form. In farming you can't take one person's success and translate it into your success." ■

From an office in Fraine Barracks in Bismarck, Major General Keith Bjerke commanded the units of the North Dakota Army National Guard, including the Air Guard in Fargo. Long active in agricultural organizations, Bjerke was narrowly defeated for North Dakota agricultural commissioner in 1988.

Jim Weinreis utilizes a home-built custom tank mixing truck for herbicide application on no-till cropland. One of seven brothers sharing an integrated farm and beef cattle operation, Jim is responsible for spraying and farming their considerable acreage. A mechanically handy bunch, the brothers modify or make their own equipment to fill specific needs on their farm.

Jim Weinreis
VVV Ranch ▪ Sentinel Butte

"Our family homesteaded in this area, and dad started farming on his own at an early age. He was also in the propane business and built the farm by buying land when it came up for sale. There are twelve kids in our family, and dad taught us all how to work and how to get things done. Except for the one year I was at college, all us boys went right to work on the farm right after high school. Now, two of my brothers run our feedlot in Nebraska and there are five of us here. I do the spraying and crop scouting, another brother heads up the seeding, one keeps the books, and the other two pretty much handle the ranching end of it. We get together to talk about things, but there's no regular meetings or anything. We think several heads are better than one and we try not to step on each other's toes much.

"I don't think I'd be interested in working this hard without my brothers. We like the family aspect of what we do. Dad encouraged us to work together. The work just seems to be continuous. Calving starts in April, then it's planting and spraying. There are always pens to clean in the feedlot and fences to fix. We're always busy.

"We raise wheat, durum, corn, barley and oats on the farm, but the biggest part of our business is feeding cattle. We market our grain or value-add our grain through our cattle. Some years it's worked. These last few years it hasn't and the financial outlook isn't promising for the coming year. Marketing is so global now. In past years some disaster would influence the market, but it doesn't seem to affect prices much anymore. The concept of just-in-time marketing by the big corporations is hurting us producers out here. All the big meat packers have their own feedlots or have tremendous numbers of cattle under their control. In a sense, they have taken away the free market because they own so much of it.

"We ship the biggest numbers of cattle to our Nebraska feedlot in February and March. We wean our calves in the fall and background them here, and we also buy cattle in order to fill our feedlots. We'll ship the heavier cattle in December to make room for our lighter, grass-fed cattle. The cattle will weigh about 800 pounds when they go to Nebraska, where we'll get the heifers to 1,100 to 1,200 pounds and steers to 1,300 pounds for market weight. The biggest percentage of the grain fed our cattle is grown right here on our farm. We also feed roughage from our irrigated corn. We sell most of the wheat and durum that we raise, but the prices are so low it's hard to recover the input costs. Now we're planting canola, peas and lentils as rotation crops on our no-till land.

"We went into no-till on a big scale about seven years ago. Actually, my dad no-tilled some of his land before it even had that name. We think no-till uses less manpower and less equipment, and it conserves whatever moisture we do get out here. Our rain sometimes comes in cloudbursts, and no-till helps control erosion. You get better use of your land with no-till. We aren't fallowing as much land, so we use chemicals to control the weeds. I think there's way more wheat being produced now because of no-till. We're producing more and getting paid less.

"We farm about seven to eight thousand acres of cropland. Some of it is rented. Rents around here vary quite a bit, from fifteen dollars to twenty-five dollars an acre. We expect an average wheat yield of thirty to thirty-five bushels per acre. A forty-bushel yield is a good wheat crop for us. We feed 4,000 head in our feedlot here on the farm.

"For us, getting bigger looked like the way to stay in business. We've made it work so far. The averages have worked out for us. When you buy equipment, it costs less per acre, so it works out there. But as farmers we're not sharing in the good economy everybody else in the United States is experiencing right now. If we looked at expanding today, we probably wouldn't do it because there's no way to make a return on the investment. It's hard for me to get over the fact that what we're producing out here is worth less every year but our input costs keep going higher.

"A well-established operation that's mostly paid for can hang on through these bad prices. Some people are hanging on, but we're seeing neighbors who've lived on their ranch all their lives getting out. It's a sad situation. There aren't hardly any family farms left anymore. It's pretty empty around here. With fewer and fewer people farming every year, who is going to raise our food? I wonder if people think much about where their food comes from? If the big corporations take over everything, we'll have twenty-dollar steaks in the grocery store because they can set the prices. Are people ready for that?

"Living out here is a way of life, I guess. You get used to it. But I think we've built a good opportunity out here for our five kids, and I'd like to see them get in the business some day. We put in long hours and just hope we get rewarded for all the work. We feel that we've been given an opportunity to use this land for a short while because it's part of God's creation. We try to be good stewards. We feel that we've been entrusted to take care of this land." ▪

Jim Weinreis is one of seven brothers who together operate a diversified grain, cattle and feedlot operation thirty-six miles south of Beach. The Weinreis brothers are fighting poor grain and beef prices by running a very large, vertically integrated operation. They are also innovative and skilled mechanics who modify or build machinery to better suit their needs.

Dr. Michelle Merwin
Veterinarian ■ Hebron

Dr. Michelle Merwin and Dr. George Amsden, a husband and wife team, opened their veterinary clinic in Hebron in 1992. A mother of four, Merwin mostly works in the clinic while Amsden handles the herd and outside ambulatory portions of their practice. The great majority of their clients are cow-calf beef producers who are experiencing consecutive years of depressed prices.

"We see the effects of low prices. Some producers will forgo treatments or preventative and vaccination programs because money is tight. Some may not pregnancy check their whole herd, they'll just have us check the questionable cows. But generally, if a critter's sick, they'll bring it in. They won't stop to think about it, they'll get that cow treated. It's part of their emotional tie with their animals.

"The prevailing thinking around here is if a producer is going to have cattle, they have the responsibility to take care of them. Ranchers accept responsibility for the health of their herd. The majority of producers around here have a hard time when a cow 's not doing well. That animal has been providing their family with an income for a long time. It's hard on ranchers to get rid of these cows. They know their animals and their personalities and how they calve and how many calves they've had. We may look at a cow as an economic unit, but for these guys there's an emotional attachment with their animals. If you've got an animal that's taken care of your family for a long time, it's hard to get her out of the herd because she's an institution.

"We become partners with producers in the care of their cattle. Most of our clients are also our good friends. We like to see them do well and obviously that helps us with our business. We want them to get ahead. The way the economy is going, there will be fewer producers, but it'll take more animals to make a living. Years ago, a producer around here with 200 animals could make a nice living. Today he needs 500 head to live the same way.

"As long as we're in town, we're on call. The work is seasonal. From mid-February to mid-May it's continuous, especially during calving season. We each probably average ninety to one hundred hours a week working with cattle, including driving and cleanup time. We cover an area forty miles east and west of Hebron, probably sixty miles north and easily seventy-five miles south of here. This is great cattle country, but there is a lot of distance. Sometimes it gets to be a long haul, especially when you get called in the middle of the night or it's cold out. But you have to be in the cold and wet if you're working with cattle. We grew up ranching and dealing with the elements and that's probably why we love it. That's why we're here. Vet medicine is as challenging and rewarding as we hoped it would be.

"Two years ago, when the weather was so bad, we saw some older producers who didn't have relatives to take over their ranch just say 'enough' and they got out. If it hadn't been for the weather and lousy markets, they probably would have stayed in another five to ten years. We see a lot of father-and-son teams around here who are doing okay, and other producers who are just hanging on. We have some ranchers who are in their mid-twenties and have just taken over the family ranch and are doing a great job. They've got energy and enthusiasm. Just about every producer we see is tied to family. There's nobody coming in and getting into ranching without a family connection to the ranch. Once in awhile you hear of a new person who bought a place because land is cheaper here than in other states, but generally people have to have family ties in order to get into ranching.

"We've added a satellite clinic in Elgin to better serve producers because of the great distances out here. We're also learning how to deal with game farms and work with producers who are diversifying into elk, bison or deer. We didn't get much of this in vet school, so we have to do research in these new areas to help producers. Bison looks like a viable industry. We've seen the number of bison grow just since we've been here. For us, bison are much more hands-off than cattle, so we work with bison only on rare occasions.

"When we first started our practice, a woman doing large animal work was a problem for some people. People would call on the phone and ask to speak to the real vet, the man vet. Our clients now know both of us can do the job, and some ask for whichever one of us they feel most comfortable with. Actually, the biggest problem is with older ranch wives who are not used to having a woman being around what's considered man's work. It's not the men who have the problems dealing with a woman; it's the women.

"We enjoy what we do. We enjoy talking with ranchers when they stop by for coffee or bring in an animal. They know we're genuinely interested in what's going on in their lives. We enjoy working with animals and consulting with producers on their care. We like the ranch lifestyle and the attitudes the people have. It's the closest we can get to ranching without actually ranching ourselves." ■

Dr. Michelle Merwin pulls a 100-pound calf from a year-old cow during Caesarian section surgery in her clinic.
In spite of consecutive years of low prices, beef producers provide the best care for their animals, she believes,
out of an emotional and economic attachment to their animals.

Gary Greff

The Enchanted Highway ▪ Regent

Gary Greff is a metal artist who is trying to draw attention to his hometown farming community by placing gigantic metal sculptures of grasshoppers, pheasants and other themes along the thirty miles of State Highway 16 from Gladstone to Regent. A former teacher, coach and school administrator, Greff grew up on a farm and often helped his brothers until they left farming. He has received national publicity in his single-handed attempt to keep his community alive.

"I've been called crazy and a lot of other things, but I don't care what people think. I've got this vision of using metal art to draw attention to Regent and I want to see it done. If we don't do anything, we're dead. Already two towns on this road, Gladstone and Lefor, have lost their schools and a lot of their Main Street businesses; and the two towns on either side of us, Mott and New England, are struggling. It's pretty obvious farming isn't doing it for us anymore.

"The Enchanted Highway gives us a chance to change things. Farms around here have gotten big, ten thousand acres and more, and that's okay because the farmers are making it. But our community is still drying up because there are fewer people and fewer kids. Years ago we had more farms with big families on each one. Today there's no new blood coming in, there's no new nothing in Regent.

"We have to change and do things differently in order to make it. There are those naysayers here who think this is a dumb idea, especially if you've lived your whole life in farming. It's hard to say we need something else to keep us going. There's so much fear of the unknown.

"People need to know there are things other than farming that can save a town. Sure, we might only become a tourist town, but you can still live on the farm and have a job at a gift shop, or a restaurant or motel. We have a co-op grocery store that dates from when the town started. Today it's sort of one-of-a-kind, but it tells you something about a town when you need a co-op in order to have a grocery store. People are beginning to realize we're dead if we don't take action ourselves.

"Ten years ago – in 1990 – I started thinking about how I could help the situation in agriculture around here. How can I help Regent? Well, what are farmers and ranchers good at? Welding. We're all pretty good welders. One guy here had built a man out of metal holding a bale of hay. He got quite a bit of publicity and people were coming here to see that one bale man. I thought, 'Imagine if we could create something more unique than one bale man?' The sculptures couldn't be normal size; no one is going to stop for normal. They'd have to be the world's largest.

"That's how I came up with the idea of creating ten gigantic metal art sculptures out on the highway. I took the idea to everyone – the city council, chamber and commercial club – and everybody was 'yes, yes, we're behind you as long as you do it.' So I've been at it for nine years now. I never wanted to have junk out there. I want it to be art. A professor at NDSU encouraged me because he thought my ideas were folk art in its finest form.

"The first idea was the tin family. I found some used oil and gas tanks for materials. The local Lions Club gave some money to buy more material and some people helped by donating things, moving tanks or helping weld. The second site was Theodore Roosevelt. Everything was donated. One farmer loaned me his forklift; another guy brought his truck to haul materials. The school loaned me their wire-feed welder. Those first two sites were more of a community effort than now.

"Next the grasshoppers and pheasants were done. The flying geese are ready to go. Each one takes one year to build and another year to install. I'm halfway there for installing a sculpture every three miles along the road. The next ones will be about deer, then maybe a Native American theme and probably something with fish.

"I'm not looking at one piece. I've always thought about a whole complex. My vision doesn't end with ten sculptures out on the highway. I see us redoing the storefronts in town with metal art, and building a golf course with metal art as a theme. We could create different colored trees for each fairway. I'd like to build it along the Cannonball River. People would come out to see that, and golfers would want to play on a course with all metal trees.

"I want to create a family atmosphere here, so families will stop on their way to Medora or the Black Hills. It's a big dream, but it's what keeps me going. Take the golf course idea — it has to be unique to draw people. We'll make it so there's not another one like it in the world. Golfers will want to come to Regent to play it, so a whole new caliber of person will come here. Once they're here, they'll see how nice and clean it is and what we've got to offer.

"Not everyone has caught my vision yet, but I see it so clearly. I've never been one to quit, no matter what I've started, so I'll see this through. It's important to me because everyone is leaving. I've got nieces and nephews and I'd like to see them be able to find something to keep them here. There's nothing here for them now. It doesn't have to be that way. I believe people influence people. I want to give people the idea that they can do something here. If you never let people see what your community has to offer, they'll never come.

"I want people to say, 'If you haven't driven to Regent, the metal art capital, and seen the Enchanted Highway, you haven't seen North Dakota!'" ■

Gary Greff believes if small towns do nothing to change, they will die in this new era of larger farms and fewer farmers. He hopes tourism, in the form of people driving the "Enchanted Highway" to see gigantic metal art sculptures, will help save Regent.

Sonia Meehl
Elevator Manager ▪ Crete

Sonia Meehl is general manager of Crete Grain, a family owned, 1.2 billion-bushel capacity elevator. Since 1997 she and her husband and two brothers have taken ownership of the elevator from her parents, who bought it from a local cooperative in 1978. She grew up on a farm near Crete and is a 1986 graduate of North Dakota State University with a degree in agricultural economics.

"Sometimes I think farm wives should do the farm marketing. They would be very effective marketers. Wives are often more astute about money. They like to get the bills paid. Farmers often get very attached to their grain and hold on to it much longer than they should. But some producers are starting to get over their attitudes. They have to. I mean, why raise the stuff if you're not going to sell it?

"Producers need to remember that once they put their grain into a bin or an elevator, they are no longer in the production business; they're in the storage business. Most producers, it seems, work for their grain bins instead of figuring out a way to make that bin work for them. A perfect example was last August during wheat harvest. The price of wheat averaged two dollars and ninety-cents per bushel, not enough to get producers excited, so they put their crop in the bin and hoped the price would go higher. But what if they had put their wheat in the bin in August and asked me what I would be paying in January? During a good deal of that time I was bidding thirty-five or forty cents higher for January grain. So if a producer had put his wheat in the bin and sold it immediately for a January delivery, he would have locked in that thirty-five or forty cents more a bushel. Plus, he might have cashed in on some premiums and discounts at the time of delivery. At harvest time I would discount as much as thirty cents for some lower protein wheat, but in January that discount would have only been about fifteen cents.

"Some people say the whole production agriculture chain will someday be linked, from a bag of seed to the chemical program to treat it, to the fertilizer and marketing. Even now, according to a contract I saw from a large company, if farmers buy their seed, they can sell them their corn at harvest and be guaranteed an average or better than average future's price. That big company is telling the producer if he buys their seed, they will market his crop for a good price. Of course, I'm not so sure farmers like to have their hands tied to only one market or a single brand of seed. I would argue that farmers could do better themselves if they got past the mindset of just holding onto grain waiting for it to go higher. It would be better, I think, if they would determine what price they want for their grain, and when it gets there, sell some.

"The whole point is, as producers and agricultural business people, we need to be flexible and do what the market wants, what the customer wants — even though that may not be what we want to do. A lot of producers don't bother to think, 'Who is my customer? Why am I doing this?' If a producer goes into the local implement dealer to buy a tractor, he's the customer. The same goes for when a farmer buys fertilizer, seed or chemicals. He's the customer, the end user. But when a producer comes to the country elevator, he needs to change his thinking from being a customer to asking, 'Who is my customer?' If producers deliver corn here, for instance, their customer is the Pacific Northwest exporter who ultimately ships their corn to Japan or Korea or somewhere in Asia. That's the real customer.

"So far that customer hasn't said, 'We don't want genetically modified corn.' So far. But that could change tomorrow. Farmers have been quick to accept genetically modified seed, but if their customers suddenly didn't want that, then what? Any business that's customer driven will try to manipulate the customer's attitude the same way advertisers do. But if the customer resists and we want to keep them as customers, then we had better adapt to what they want.

"The country elevator is really just a go-between with the producer and customer. We gather up large volumes of grain and blend it to the specifications our customers want, then load the grain into railroad cars for shipping. This provides producers with access to customers who might be millers, exporters, processors or feed mill operators - whoever makes the commodity into something consumers want.

"Providing access to customers has changed a great deal since the days of the old country elevator. As railroads have become more efficient, rate structures and incentives have encouraged larger trains and faster loading. Until twenty years ago, a train would generally move along the line picking up individual boxcars from many elevators. Railroads then starting running unit trains of twenty-seven or fifty-four covered hopper cars; usually all loaded at the same elevator. We're a unit train loader on the Red River Valley and Western Railroad. Typically, after grain cars are delivered for unit trains – I have fifty-four empties sitting out on the siding right now – we have one day to load the train and take grain samples. From here the train takes cars to Breckenridge, Minn., where they are sorted for the Burlington Northern Santa Fe to continue in unit trains to the West Coast, Duluth, Minneapolis or points east. This change hasn't been an easy one, judging by the abandoned country elevators all across the state. The rate difference between a single car and a unit train may run ten to fifteen cents per bushel, which is too competitive for the old-time, single-car loading elevators.

"Twenty years ago, fifty-four car unit trains were a big deal, but now railroads are moving toward unit trains of seventy-five cars and even100-plus cars. The railroads have been giving us at least twenty-four hours to load unit trains, but with lower rates for longer trains, they would give us only fifteen hours to load a 100-car train. You can't load that volume – 440,000 bushels of corn, for example – that quickly at a facility like this. So we've had to start thinking about expanding to accommodate 100-car unit trains. Just like producers, we have to be flexible and do what the market and the customer wants – even though it may not be what we want to do." ▪

Sonia Meehl stands on a railroad track siding where she is contemplating building a one hundred-car grain loading facility. Just as farmers must change to survive in new economic realities, Meehl believes country elevators must invest in themselves to best serve their customers.

Jon Wiltse believes what is being discarded by most farmers can become an economic advantage for struggling farmers. With an abundance of grain being grown in the region, the opportunity to utilize straw could put money into the pockets of farmers and create jobs in the region. The challenge, one that continually plagues new business ventures in the Dakotas, is finding access to sufficient capital.

Jon Wiltse
Investing in Straw ▪ Lisbon

"Farmers today need to think of ways of adding more dollars per acre. One way of doing that is selling the straw from wheat fields. We're doing this in the Lisbon area. Harvest Board International is getting ready to start making strawboard here. They know that with the big lumbering operations running out of timber, there's a good market for strawboard. Strawboard is thirty percent stronger than wood particleboard - and twenty percent lighter. Strawboard also holds screws better and it's more water-resistant. Oddly enough, strawboard isn't cheap; it's pretty close to the price of wood.

"The strawboard company wants local people to provide the straw supply and bring it to their plant. Heaven knows there's plenty of straw out here — we figure over a million acres of wheat are planted within a sixty miles radius of here, probably enough to supply three plants like the one we'll be having at Lisbon. I got some farmers together and that was the beginning of our co-op. We're expecting to make ten dollars a ton on our wheat straw, minus about a buck per ton that will go to running the co-op.

"Our co-op is basically a service co-op in which everybody who wants to be in it contributes some money to pay the expenses and obtain the straw supply. If you want to sell straw, you become a member of the co-op and you buy shares for the tons of straw you want to deliver. We generally figure a yield of a ton of straw per acre. So if I had 1,000-acres of wheat and wanted to put 1,000 tons of straw into this plant, it would probably cost me two dollars per ton, or $2,000, to get into the co-op. Since we'll be drawing from such a wide area, we'll share in the transportation costs so that the guy living fifty miles away gets the same price as the person living five miles away.

"The co-op will go out and get the straw from the farmer and then deliver it to the plant. The plant will then pay the co-op, and the co-op will give the money back to the farmer and keep whatever little money it needs for operations. You might say the co-op is kind of the middleman between the farmers and the strawboard plant.

"For farmers, this is a good way to add value to our land, a lot better, I think, than farming more land. Being bigger in times like these is just about guaranteeing yourself bigger losses, especially with the low prices they're predicting and with the profit margins so much thinner than they used to be. I suppose that's why we sometimes hear older farmers wonder why those of us in the younger generation can't make it in farming. I don't think they realize how much tighter the profit margins are and how much faster the inputs go up than the prices we get for what we produce. I remem-ber around 1980 when my dad retired and moved to town and he still had all his tax records, back from when he first started farming. It seems to me that for every dollar of expense, he had maybe six dollars of income. Today, if you spend six dollars and get seven back, you think you have a pretty good profit margin.

"Several years ago I was telling other farmers it was time to get out of farming. Get out, and then maybe get back in after three or four years. But I didn't take my own advice. Why? There are several reasons why people are reluctant to get out of farming. Family, for one thing. I farm with my brother, and we're a fifty-fifty partnership. For others, farming's their life and they can't imagine doing anything else. Some can't imagine they have the ability to learn to do anything else.

"To survive on the farm these days it's not enough to use just your back; you have to use your brain. You have to become a businessperson. If you can't do that, you're going to go broke. That's almost a guarantee." ▪

Jon Wiltse and his brother Dan are the third generation on their family farm east of Lisbon. Wiltse is helping to establish a wheat straw cooperative to supply a strawboard manufacturing plant being planned in the Lisbon area.

Fred Kirschenmann

Advocate for Sustainable Agriculture ■ Windsor

Frederick Kirschenmann earned a doctorate in philosophy from the University of Chicago and in 1976 returned to the family farm to convert the 3,100-acre grain and livestock operation into an organic farm. He has published numerous articles on sustainable agriculture, and his farming philosophy was featured in the award-winning video "My Father's Garden."

"There are three essential problems that farmers have to start working on if we want to get out of this ongoing farm crisis. One is to internally develop production systems that rely more on management than on inputs. The more we can internalize our costs, the better control we will have of them. It doesn't mean that costs are necessarily going to be lower, but farmers will have control of their costs.

"For instance, if the cost of nitrogen doubles, as it did here a few years ago with anhydrous ammonia, we have no choice but to pay it. But the cost of livestock manure and yellow blossom sweet clover doesn't change much. We would have control of our costs with those sources of fertilizer.

"Second, I think we need to look at recapturing part of the market share. This is what Dakota Growers Pasta Cooperative has done. Now, that's certainly not a risk-free venture. One of the things I'm very pleased about in North Dakota is the way we've approached these ideas. There's some public money available so farmers can pool their resources and get grant money to do feasibility studies and business plans to reduce their risk.

"The third thing is, and it's something none of us has done very much about, we have to increase our collective bargaining position as farmers. Farmers have virtually no say either at the table of government or at the table of the marketplace. There is a lot of money in the food system. The problem is, farmers can't get their fair share of the pie. Ultimately we are going to have to do something about this.

"The irony is that we are already paying "union dues" in the form of check offs — more than two billion dollars a year in check-off programs. Very few of the check-off benefits come back directly to farmers. What is the benefit? The check-off dollar is used to try and get people to eat more potatoes, or eat more corn or wheat or drink more milk. But even if farmers are wildly successful in getting people to consume more, the increased sales won't benefit us much because our share of the income from increased consumption is so little. Our two billion in check-off dollars is only earning more money for the market sector that has already taken the lion's share out of the income pie.

"There are things we could do if we ever get farmers to really start seriously talking to one another. We could simply say, 'Let's pool the money.' If we put all our check-off dollars together and democratically elected representatives to really look after the interest of farmers, I think we could have a real bargaining position and get our fair share out of the markets. As farmers we have to find a way to look out for ourselves.

"I was teaching at the university when I became aware that one could improve soil quality with alternative methods. We have all of this land out here entrusted to our family, and it started to bother me that we weren't doing the best job we could to maintain the quality of the soil. So when I came back to farm, I decided that I wanted to try and improve it. It didn't take a rocket scientist to figure out we were using increasing amounts of fertilizer in order to maintain the same yields, so we were doing something to the soil that was causing it to no longer be as productive as it once was.

"A cornerstone of sustainable agriculture is to develop nutrient recycling systems inside the farm instead of relying primarily on inputs from outside the farm, and to develop self-regulatory systems like predator-pest relationships to control insects. What we're doing by not paying attention to these whole-system approaches is creating agriculture that is more costly and is generally more damaging to the environment. Probably one the most obvious examples of self-regulatory systems are ladybugs and aphids. If you've got a good crop of ladybugs out there, you never have to worry about aphids in the fields. This is what I mean by a self-regulatory system. If you don't have ladybugs, then you have to get on a tractor and apply an expensive chemical input to get rid of the aphids.

"If you look at some of the contracts that farmers have to sign now if they buy genetically engineered varieties of seed from say, Monsanto, you will see by the contract the farmer doesn't own the seed or the crop. Monsanto retains ownership. So Monsanto not only sells the input to the farmer, they also retain ownership of the crop the farmer produces. What the farmer is basically doing is contracting for permission to produce Monsanto's crops for them. So we've come to the point where farmers are no longer farmers as we have understood them in the traditional sense.

"We have a whole range of questions we need to ask ourselves now as a society. Do we want our entire food system in the hands of a half dozen transnational corporations? Historically, these corporations have no allegiance to people or to country. They only have allegiance to the bottom line. We're really at a point now where farmers are rapidly moving into a future where we have no freedom. One of the reasons why people have traditionally wanted to become farmers is the freedom to run their own operation. But we're being locked into a contract system that gives us no incentive or freedom to make choices that are in the best interests of the farm, let alone the best interests of society. I think that's a serious issue today. It's not very well understood that this is where we're rapidly heading in agriculture." ■

Fred Kirschenmann stands chest high in one of his fields of grain that will bring a premium price because it was raised as certified organic. He advocates using as many on-farm inputs as possible to reduce overhead.

Dennis Kubischta
Flour Miller ▪ Hope

Dennis Kubischta farms 1,900-acres and mills his own wheat into flour that he sells commercially. He has proposed building a mill as part of the empowerment zone for rural economic renewal proposed by the federal government in Griggs and Steele counties.

"When I talk about diversifying, I don't mean if you raise wheat and barley, maybe you should raise corn and soybeans, too. That way you're still locked into the commodity market. To diversify, I mean to somehow use your farm to value-add to the commodities you already raise. There are not many farmers around here diversifying. They might have some ideas, but where is the money going to come from? There isn't money or programs out here to diversify these farms. That's one reason why many operators are quitting rather than diversifying.

"I've talked with lots of farmers who have quit farming. Good farmers, too. I've not talked to one who regrets quitting. My great-grandfather came here 100 years ago and to be very honest, it won't be easy for me to quit. Giving up is incomprehensible to me. Maybe it relates to Vietnam and the soldier's code of conduct never to surrender of my own free will. I feel I've still got the means to resist. I have a plan.

"Since I've been milling my own grain, I've kept some of the money that would have gone to the middlemen. If I could do it on a larger scale, I could do even better. I started ten or twelve years ago when I got into sheep. I knew I needed to diversify; I was farming only grain. I was growing forage and feed grains already, so I thought I could better utilize our resources with sheep. I bought a cleaning mill at the same time so I could clean grain and feed the screenings to the sheep. That's the way I started value-adding our grain.

"A short while later I found a flour mill for sale. It was everything you need to make whole wheat flour. I got a loan to buy the equipment and started milling to the standards of the State Health Department. Today I deliver flour from the farm to Quality Bakery and Hornbacher's bakeries in Fargo. These are good accounts; they use a lot of flour. If you go to the Fargo Holiday Inn or the Radisson and you eat any whole-wheat products, it's my flour. It comes from my wheat and my land. A distributor from Fargo has begun selling our flour in stores from Alexandria, Minnesota, to Mandan. We bag all the two-pound and five-pound bags right here on the farm.

"One of the reasons I was confident about milling my wheat was the Top Taste Bakery in Finley. When I asked them about buying my flour, they said if it matched their standards, they would buy from me, and the standards for a frozen bread bakery are very strict. My first 500-pound batch for them wasn't fine enough. But I got some help from North Dakota State University on what size screens to use and I've been selling to Top Taste ever since. This is one good way to add value to my grain because I've eliminated the middleman.

"When I found out how large a user of flour Top Taste was, I asked if they would buy white flower from me. They said they would if the quality was there. But the cost of making quality white flour was prohibitive. Now the empowerment zone funds from the U.S. Department of Agriculture may help. I'm working with the bank in Finley for the $750,000 needed for a quality mill.

"The mill will be in Finley. The bakery will build a new building in the industrial park and the new mill will be right beside it. Top Taste needs about fifty tons of white flour a week, and being side by side, we'll eliminate transportation costs. Those savings can be passed along to the farmers who sell wheat to the mill.

"As soon as this plan gets going, I'll be way beyond my capacity to use my own grain in the mill. I plan on proposing paying farmers a forty to fifty-cent a bushel premium if they will raise the wheat varieties I want. Grandin wheat is best for milling flour for baking. Unlike the big mills that mix all the grain together, I keep the varieties separate. I can tell the difference. I intend to make all types of flour – white, enriched, rye, barley, oats, whole-wheat, everything.

"To be a good farmer, you have to be somebody who doesn't give up. You have to see some future beyond the immediate year you're dealing with and make some plans." ▪

Dennis Kubischta has won awards and praise for the high-quality flour he mills from wheat grown on his farm.
With sufficient capital - an illusive goal - he dreams of linking a flourmill with the Top Taste Bakery in Finley to provide
value-added opportunities for area farmers.

Mike Degn

Positive Mental Attitude ■ Sidney, MT

Mike Degn farms both sides of the Montana-North Dakota border in the Yellowstone River valley. A "city kid from Billings," Degn started out in 1979 with a half-ton pickup truck and his dad's 600-plus acres near Sidney. He has since built his operation into a diversified farm of 6,000 irrigated acres, 18,000-acres of dryland, a cow-calf and dude ranch, a Buckmasters hunting lodge and most recently, Moo Juice Dairy, a large commercial dairy operation.

"Right now is the most positive time to be in farming — when it's as bad as it gets. When it gets as good as it gets, that's when you don't buy any machinery or you don't buy your neighbor's farm. When times are good, you know you're headed back into the hole again. You'd better have things paid off or getting paid off before you head back down the cycle.

"Farming is always cyclical. Everything we do out here has ups and downs. Right now agriculture is on the low end. Actually, we should be happy about that because commodities always come up again once they hit bottom, and we've already achieved bottom on grain. Prices are coming back on the livestock end, so I think we're headed toward some good times. This is a time to be optimistic about farming. It's a good time to buy farms and buy equipment at good prices. People will deal with you now because nothing has been selling for a long time.

"I was such a poor risk when I started nobody would lend me any money. I only needed $5,000 that first year to operate. I worked hard and had a good year. We just made it with a lot of luck and running into a whole lot of nice people. This area is just covered with fantastic, nice people. I'd pick up a little advice here, a little there. You're never done learning. Every year we do something different. This operation has never stayed the same. Every year we re-evaluate what we do and we're always looking for something better.

"If I have a secret, it's probably enjoyment in what I'm doing. You can put up with a lot of pain and hardship as long as your attitude is good. You can work through any problem. Every day out here on the farm there are going to be problems and if you let them bug you and get to you, you're going to quit. The other trick is to be such a darn nice guy that everybody likes you and wants to work for you. There are so many wonderful people who live around here and like being here. We genuinely like each other. It's nice when your employees get just as excited as you do about the harvest. They have pride in their job and they feel a part of the organization. I'm totally relying on these guys to do the best job they can. I don't have much control over what goes on in the field anymore, but I've got good equipment operators and good mechanics who care about the job they do. The whole key to getting bigger is giving your employees a sense of belonging. It's the relationships you build that matter. The key to any successful operation is a happy workforce. A lot of times it has nothing to do with money. It's people taking pride in what they are doing.

"I'm diversified because I like farming so much. I want to grow every kind of crop I can possibly grow around here. I also want a monthly income. With sugar beets, it's eighteen months from the time you plant the crop until you get paid. That's why I like my sweet corn operation. I'm getting cash sales when nothing else is bringing anything in.

"Moo Juice Dairy has a lot to do with Holly Sugar wanting a three-year rotation on irrigated beet ground. We're going to have 40,000-acres around here that needs another high value crop. Sugar beets were so good for so long, everybody got lazy and went to a two-year rotation and lived off the beets. But the margins are tightening up and its getting a lot less profitable to raise beets now. We'll raise corn and alfalfa in rotation with beets and feed dairy cattle.

"Starting a dairy probably isn't the smartest move now because milk prices are low and prices for dairy cattle are high. It's a little scary because you never want to buy in on the high side. But getting the dairy going stabilizes my corn ground, my alfalfa crop and my sugarbeet rotation. It's just unfortunate timing. I'm working with six equal-size investors. We'll truck the milk out of here with our own trucks and tanks. It will help keep my employees busy the year around. I've got my neck stuck out a mile. I've expanded more in this low cycle than I ever have. But it's a calculated risk I'm almost positive will work. I'm acquiring more land and buying machinery cheaper than I ever have, and when we come back out the other side, this should be a highly profitable operation.

"It never hurts to be optimistic. I've never seen a time when you couldn't get out of something bad by working on the problem. But if your attitude goes bad, you're done. When farmers are struggling and hate what they're doing, it tears everything – the farm, the family – apart. I've decided that what causes farms to fall apart is mental attitude. There's no doubt it's caused by financial stress, but maintaining a good attitude is the most important thing for staying in farming.

"For me, it's fun to put it all together. I like trying different things and making them work. I like farming and I like working the year around. There's always a challenge. A lot of it is a lack of fear. I don't have any fear of failing. I truly believe I cannot fail. That scares the bankers, but having never failed at anything and having an attitude that's positive, I don't think failure is possible. What could happen out here that I couldn't work through? What problem couldn't be solved by working at it? Part of why I enjoy this so much is because I've been a city kid and lived in town. The contentment I get from farming is knowing what city life is like, and I like farming better." ■

Mike Degn says he likes to raise anything that will grow on the sizeable acreage he farms on both sides of the Montana-North Dakota stateline. He successfully raises carrots that he feeds his cattle until he can penetrate markets. In the background is organically grown sweet corn, a crop that brings a premium price in select markets he serves by running a fleet of refrigerated trucks.

The profit in grain is with the big millers and retailers, so for farmers like the Solbergs who are losing money raising wheat, there is an absurd irony in feeding sheep free day-old bread given out by local grocery stores.

Warren and Mara Solberg
Diversifying Through Education ▪ Wild Rice

"We both grew up on farms, but we didn't see many changes back then. We raised wheat and sunflowers, we had cows. There was never any thought of, well, let's do something else, like start a truck farm. But today, if farmers are going to stay on the farm, we have to be ready to change.

"Farming is our big love. We like the farm life, but we could see that we would eventually have to leave the farm and go off and work. There isn't any money in raising small grains anymore. So we asked ourselves, if Mara is going to spend the rest of her life working off the farm, why shouldn't she go to college and get a degree and do something she would love to do? Anyway, we've noticed if parents have attended college, their children usually do, too. It's so difficult to make a living today without a college education.

"Mara is majoring in psychology at Concordia College and I've taken a job in maintenance at the college. Mara enjoys college. She's found it's about exploring ideas and forming opinions. So many of her classes ring true because of the life experiences she's had. Mara has been proud of being a farm wife and a stay-at-home mom, but she's looking forward to earning her degree. To farm today you have to diversify and plan ahead. Our whole family supports Mara getting a degree because we're all in this together.

"When the Solbergs came over from Norway in the 1800s, this is the tract of land they settled on. So growing up here, we feel attached to the land. Our family has been here before us and we'd like to pass it along to our children. There's a graveyard south of here with some of our ancestors buried in it. Some day we hope our children will live here, or at least one of them.

"It used to be every year we'd go in to the bank and get an operating loan. It was merely a handshake and sign here and here's the money and go ahead and farm — come back when you can make the payments. Now, you don't know how humiliating it is when you have to go in with your hat in your hand. You can just see it in their eyes. So many personal questions and you don't sense their support like before. It frightens you that there's people out there that you thought were your friends and willing to help you, and then it changes. Now it's just business.

"We've seen farmers around us lose their farms, lose everything. Many of them have farmed their assets away. We decided it would be healthier for us to make the decision ourselves to get out before it's too late, before it comes to an auction sale. We don't want people knocking on our door telling us to get out. We looked at last year's low grain prices and thought there would be too many cards stacked against us this year.

"We've given up all the ground we were renting from neighbors. Our plan now is to farm eighty acres. Put a crop in and hopefully keep the equipment, maybe do some custom work, and then pay off our term FHA operating loan. If we can get a lower interest rate and if we can get it put out over five or seven years, or even longer, we can probably keep our equipment.

"Five years ago we got into sheep. They eat everything. At times they would just as soon eat a thistle than walk a quarter mile and eat good grass. The wool market isn't that great, but we still shear the sheep. We've done okay selling lambs for slaughter directly to people. For Mara's birthday this year a friend of ours gave her some baby chicks. We can just hear our aunts who worked hard on the farm all their lives saying, 'Why would you want to raise chickens?'

"We go to a place in Fargo that gives away food to low-income people. They don't want to waste leftover bread from the grocery stores, so they give it to us and we end up bringing it out here and feeding it to our sheep. For them it's like getting candy. Do you suppose our grandparents would ever have believed the price of wheat would drop so low that perfectly good bread would become free feed for sheep?" ▪

Warren and Mara Solberg have farmed since the mid-1970s on land that has been in Warren's family for four generations. Worried about the future of agriculture, the Solbergs are preparing themselves for a life in which farming may not provide the kind of living they would like to pass on to their three children.

Mark Willows
Aquaculture ■ Binford

Mark Willows is a cattle producer who began fish farming as a way of diversifying his farm and increasing his family's income. In late 1999 the combination of low prices and high expansion costs forced him to quit raising fish and become marketing director for the North American Fish Farmers Cooperative, which represents five producers in this region.

"With fish farming you get into the same problem the rest of the agricultural industry is facing – you're competing with everybody else who is trying to sell for the lowest cost possible. Problem is, the cost of production doesn't go down for the small guy.

"When we started, my wife and I thought fish would be a small family farm thing where we could add $50,000 to our income. We borrowed a lot of money to do it. We risked everything we had. Now we're out of the fish-raising business, but marketing was taking up more and more of my time anyway, so it just seemed logical to go in that direction.

"People say tilapia was the fish Christ fed to the multitudes. It looks like a perch, with yellow markings, but it will grow a lot bigger than a perch, more like a bass. It's a warm water fish that originated in Africa and has been aquacultured for thousands of years. Tilapia are vegetarians; they eat plant proteins such as soybeans, corn and wheat – a good match for what we grow in North Dakota.

"In some states, fisheries are cutting back on production of fish that are important commercially, so the supply of fish is shrinking at the same time more people are eating fish for its nutritional value. I see aquaculture providing most of the seafood that will be consumed in the U.S. in not too long a time, maybe twenty years. Now there's about eighteen million pounds being raised each year in the United States.

"Ninety-nine percent of the domestic production of tilapia being raised in the U.S. today is sold as live fish. In North Dakota we have to fish farm indoors, which means a well-insulated, moisture-proof building with water tanks, filters, circulation pumps and oxygen equipment. In other words, plenty of overhead.

"We haul about 10,000 pounds to market at a time, mostly to the big cities on the East Coast and Canada. The markets are best in cities where there is a big "China town," like in New York City. We use a flatbed trailer with tanks. From here it's eighteen hours to Toronto and twenty-four hours to New York City. We'll generally lose only a handful of fish from the whole load. Shipping live fish to New York or Toronto when it's thirty below gets to be a challenge, but we've done it for eight years now.

"Feed and electricity are the two major input costs. You heat the water with whatever is most economical in your area. You could use waste heat from an electrical power plant or a sugar-beet plant. Some smaller systems use coal-fired boilers or natural gas. What we've been finding is the closer you get to the markets in the more highly populated areas, the higher the cost of production. Rules and regulations are much stricter on the East Coast. And then there's the cost of labor. So what they gain on our disadvantages with freight, we gain on their cost of production.

"When I started in 1991 the price for tilapia at the farm was about a dollar and a quarter per pound. The cost of production was roughly a dollar, so we weren't making a whole lot of money. As the years progressed, the tilapia market got progressively better and in April 1998, we were selling tilapia for a dollar eighty a pound at the farm. So we were seeing some good money.

"The big corporations saw that, too, and they figured out how to do this on a much larger scale for a lot less money per pound. Rather than raise 100,000 pounds a year, they raise three million pounds. They don't have to get forty-cents a pound, they can get fifteen-cents and still make money. Also, catfish farmers down south converted part or all of their production to tilapia. In Third World Countries costs run cheaper because of subsidies. In some countries, the government even builds the processing facilities, so these countries are shipping fish into the United States for at least forty cents a pound cheaper than we can raise it.

"Right now, we're looking at selling live fish at eighty-to-ninety-five-cents a pound, and our cost of production is still up around eighty-cents to a dollar and twenty-cents a pound, depending on capital investment and input costs. We have a lot of fish farms that are losing money.

"What we are producing has become a big commodity. I guess it's no different from grain farming. Big corporations are going back to the bonanza farms. Unless something changes, you're going to have corporations owning everything and the traditional farmers are going to be working for them." ■

Mark Willows has discontinued his own fish farming operation to concentrate on marketing tilapia fish to East Coast markets. Even with transportation costs added on, fish farming in North Dakota can be competitive because of low production costs and a reputation for quality.

A professionally trained artist, Bill Lowman illustrates his books of cowboy poetry with his own drawings. In recent years his poetry has provided the cash flow for his ranching operation.

Bill Lowman

Rancher, Cowboy Poet, Artist ■ Sentinel Butte

"We ranch eight sections of privately-owned land and run 200 head of cows in a cow-calf operation. We don't dare run any more than that for fear of overgrazing these fragile grasslands. We have about 450-acres we use for a feed base and put up hay, and no, we can't make a living on 200 head. Used to be, you could live on 100 head. We'd probably need sixteen sections of land and at least 350 head to make a living just ranching.

"We knew early on we'd need to diversify in order to stay living here. Cattle prices are always up and down. We live thirty miles from the nearest town, so it isn't practical for JoAnn to work outside the home. I ran heavy equipment for years building roads. I also dug stock ponds for ranchers, but it was mainly road building wherever there was work. Even when credit was easy to get in the 1970s, we resisted the chance to borrow. We learned to diversify to survive. An old timer once gave me some sound advice. He said, `It's not the hard times, it's the good times that will hurt you.'

"We used to have Welsh black cattle, but the buyers said they were too coarse. Now we have Angus bred with Charolais bulls. We sell our calves either at the sales ring or to a feed lot operation. All these other jobs we do mainly to stay on the ranch. It's our family ranch and that's very important to us. Out here we have to depend on people for help and we know we can depend on our family.

"The cowboy poetry happened naturally. I've always done it. I'd be out working, you know, fixing fence, and get to thinking. Pretty soon I'd have a story going in my head. When I got home at night I'd write it down. Sometimes I illustrate a poem with a drawing. When they were looking for cowboy poets, the people in Elko, Nevada, found me. That first gathering really helped get cowboy poetry going. Now I do about forty banquets a year. I've got four books of poetry and cartoons out, and the banquets help sell the books. I've kinda got the tradition of cowboy poetry going now and it's been stable. Those forty banquets a year are our cash flow for the ranch. They put groceries on our table and make the payments on our equipment and vehicles. My poetry is based on my ranching experiences. It's a business for me now.

"It's very important to us to be able to live here. We are surrounded by all this beauty and being outside in nature is a blessing. We thought other people would like to enjoy this too, so last year we opened Lowman's Wannigan Creek Lodge. It'll sleep eight and it has a nice kitchen and bathroom. The main thrust is for mule deer archery hunters. One of my sons and I built the cabin from logs we bought from a friend in the Black Hills. Some of our guests say it's too nice for hunters. We like hearing that because we want to do it right and make this a nice place to stay. We've been approved for as many as ten cabins, but we're going to take it slow and only grow to meet demand. We had a family from Iowa stay here this summer because they wanted a true Bad Lands experience. We were featured in the L.L. Bean outdoor catalog this year and we hope that will be good exposure for us.

"The lodge is working for us, but it takes work. We have to promote it. Our hope is the lodge will take over for the construction work. I'm getting too old and beat up for construction work. Too many horse wrecks in my wild and reckless youth. We have good hunting here, and there's great scenery. That Iowa couple just wanted to hike and enjoy the country. We offer a good experience in a good facility. Our challenge is to find clients for that first visit. After that we think they'll come back again and tell their friends to come out.

"There's very few people who live out here. We like the solitude, but it bothers us that there's fewer people living out here every year. The cattle industry is staying steady. It's poor, but not a complete disaster. With our experiences, we doubt people can start up ranching from scratch. It's too expensive with the returns as poor as they are. We think we can hang on. We've always been diversified. If one business doesn't work out, we'll start another. With the lodge now, and the poetry being steady, we feel confident we've put the pieces together. This is where we want to live and we'll do whatever we can to stay here." ■

The telephone answering machine at Bill and JoAnn Lowman's Wannigan Creek Ranch north of Theodore Roosevelt National Park asks callers to leave a message for one of four businesses they operate in order to stay on the ranch: their cattle operation, a hunting lodge, heavy equipment work and cowboy poetry speaking engagements.

Pick My Own Friends

As ah cowboy on earth
as each day begins
I reserve the right
to pick my own friends

The right to be choosy
the right to be free
The right to be different
if I so choose to be

I'll judge the horse
I want between my knees

And I'll pick my own friends
just any I please

There's some I don't care for
that's my business too
You'll never catch me
tellin' you what to do

It's not what you are
that I choose to be host
Rather how you are
that matters the most

Dennis Sexhus stands among bison carcasses in the New Rockford processing plant. Sexhus is proud of the fact that members of the bison cooperative are some of the only farmers and ranchers who are able to influence the price of what they produce.

Dennis Sexhus

North American Bison Cooperative ▪ New Rockford

"We are in a completely unique situation. In most markets, the price of commodities and livestock is determined by the market, and that price changes by the day. But we're different. Since 1994 our members have set the price for animals they deliver here for slaughter. And that price – $2.35 per pound – hasn't changed in six years.

"The value-adders are the only ones making money in agriculture today. We feel that as producers we can control the supply, so we can influence the price. We're kind of blessed that we are in an industry that's small enough that we can exercise some degree of control over our fate. The North American Bison Cooperative accounts for about sixty-percent of the bison that are slaughtered in the world and we think within another year or two we will have seventy-five percent. We have come to the point now where we influence the prices for virtually the entire bison market.

"The people who own this cooperative have decided to sell meat instead of animals. I like to think of it as controlling our product from gate to plate. This is a lot different from the way farmers have generally thought of themselves. When I was growing up on the farm, I thought when we loaded our stock on the trailer or when our wheat was on the grain truck, our job was done. We thought of ourselves only as producers, and we got our check at the time we delivered our animals to the sales ring or our grain to the elevator.

"If you buy a share of stock in the cooperative and deliver an animal here, you get your check when the meat is sold, not when the animal is butchered. It's a concept I think agriculture is going to have to accept. If we're going to get good returns, we'll have to look beyond the farm gate.

"The North American Bison Cooperative is a closed cooperative, limited to producers, and we slaughter animals only for our members. This makes it possible to guarantee members a price that allows them a fair return for their work and their investment, as well as a market for their product. Our rules require all members own a minimum of ten shares of stock, which means all members are required to deliver at least ten bison a year for slaughter. The price of a share is six hundred dollars. Now, some guys have a lot of shares and a lot of buffalo, like Ted Turner. He's a member and he comes to our annual meetings, and he ships in a lot of animals. I think his heart is in the right place. He has a concern for the land and he thinks bringing back buffalo is a good thing. But each member only has one vote, regardless of the number of shares he owns.

"Our strategy now is to market primarily to white-table cloth restaurants in metropolitan areas where low-fat, highly nutritious bison meat is attractive to health-conscious peo-ple. We are almost one hundred percent focused on selling to top restaurants where people are willing to try a new product. There are other reasons, too. A top restaurant will have a chef who will cook a bison steak right. If someone gets a bad bison steak, they'll probably never buy another one. We've all had a bad beefsteak, but we still order it because we know it's good. The second thing is, by the time enough people have tried buffalo in restaurants, we'll have sufficient product to stock grocery stores.

"We've started a national marketing campaign. Each member is required to contribute seven percent of the carcass value or approximately one hundred dollars per animal for marketing. We'll spend about $2.5 million dollars this year on our program. Our subsidiary, North American Provisioner in Omaha, handles all of our distribution. We host seminars to train chefs how to prepare buffalo and we go to industry trade shows. The jury is still out on how well we'll do, but we believe we're on the right track to market bison nationally and internationally.

"We also sell the by-products. Just like the Native Americans of the past, our coop ensures that every portion of the bison is used for high-quality products. We sell the skulls, hides, horns, tails, everything. The hides are sold to a boot and shoe manufacturer; the skulls are in high demand. Some are decorated by Native Americans and sold in bou-tiques in places like Santa Fe and Jackson Hole.

"Our original challenge was to sell meat; now our chal-lenge is finding enough bison to fill the orders. Our industry is growing at a twenty- percent compound rate. In the last twelve months we've grown from 250 to 350 members and from slaughtering 7,500 head to 9,500 head a year. Bison are already the second most important livestock industry in North Dakota. We're building a second plant in western Canada, and we're planning two new additional plants. Our goal is to service all the producers in North America with plants that are within reasonable driving distance.

"I think our cooperative proves that these farmer-owned businesses can work. In our case, as our co-op goes, so goes our industry. What makes this work is we're selling an all natural product that tastes good." ▪

Dennis Sexhus grew up on a grain farm near Leeds, and since 1994 he has been the chief executive officer at the only producer-owned, buffalo processing and marketing cooperative in the world. Sexhus previously spent twenty-eight years in senior management with several manufac-turing companies in the United States, Canada and Europe.

Lyle McLain believes the economic future for elk ranching is bright. Until adequate supplies are available for the butcher market, producers can sell antlers to the Asian market. Antlers are ground into medicines to treat cancer, heart disease and arthritis.

Lyle McLain
Elk Rancher ▪ Mohall

"For us, it got so that cattle didn't make economic sense. It seemed in one year out of seven prices were good; in the other years, they were either on the way up, or they were on the way down.

"We got into elk in the fall of 1991. I went with some other elk producers to a sale in Minot that was being conducted by satellite from Houston, Texas. I sat there and watched it on TV. Before it was over, I got up, went to the phone, and bought five heifer calves – my first elk.

"So we had the elk, but we didn't have a fence to keep them in. The state requires the fence to be a minimum height of eighty-inches. Fortunately, a fellow who lives about ten miles west of here that had been raising elk came over and helped us put up a fence. We fenced in about three-and-one-half acres.

"You might think elk are hard to keep confined, but we've found they're not. Elk are very territorial. They tend to stake out their territory and make that territory their fence line. Once in awhile, when somebody is careless about closing a gate, we'll find an elk outside the fence, trying to figure out how to get back in. Bulls are more territorial than cows. During the breeding season, if you have a bull in the pasture with eight or ten cows, there's no way you can chase that bull out of that pasture.

"Elk are a lot different than cattle. If you think you can go out and chase elk from one pasture to another, the way you would cattle, forget it. It just doesn't work. All you need to do is open the gate and just watch, and probably in ten minutes the elk will notice something is different about their territory - the gate is open where it was shut before. And just for curiosity they'll go and investigate it. There is always one leader who will take the herd through, and it's usually a female.

"At every feed seminar I've been to, we've been told not to ruin elk the way we've ruined the beef cow. The Hereford, when it was first bred and registered, was 800-pounds of good quality cow. Now we feed them up to 1,700-pounds and they've lost much of the characteristics of their original breeding. The worst thing you can do to an elk cow is let her get too fat, especially at calving time, which is May and June, because you'll likely lose the calf and the cow both. In the first part of February, I get my elk cows started on a program of losing weight.

"You don't feed elk the way you feed cattle. I feed mine oats with a supplemental mineral pellet. Some farmers feed corn. Elk are grazers and will consume as much as twenty percent of their diet in browse.

"Elk meat tastes good. It's lean and nutritious. It appeals to a lot of health-conscious people. But not much elk is sold for meat in North Dakota for one simple reason: there aren't enough elk for slaughter yet without destroying the breeding herds. There's a guy in Montana with a butchering facility that supplies meat for about 120 stores in six states. Two years ago he claimed he could dispose of 2,000 head of elk a year through his marketing. At that time there were, if you counted every elk in captivity in North Dakota, only about 1,200 animals. So you see what I mean.

"So far the chief economic benefits of elk have been breeding and antlers. For over 2,000 years Asians have used antlers not only from elk, but also from other antlered animals, like reindeer. The main use we've heard about in this country has been as an aphrodisiac, but some Asians believe the antlers prevent cancer, heart disease and arthritis. In fact, there are fifty-three different ailments they believe benefit from elk antler. In my case, I had arthritis so bad I could hardly walk. Two years ago I started taking pills made from elk antlers. I'm not ready to join the ranks of the marathon-runners, but I can do an eight-hour job now where I couldn't walk before.

"Because of the Asian economy, the market has recently dropped off. But we expect things to improve as the Asian economies improve. Asia has the largest population in the world, and it's the fastest growing, too. They still believe more in natural cures as the way to good health than in prescription-type medications. Those cultures have emphasized natural remedies for thousands of years, and we don't think that's going to change anytime soon.

"Even in the United States, elk pills are starting to show up in health food stores. There's a chain of about seventy stores on the East Coast that are going to be offering them, and we expect the demand to increase. Supply is going to be the problem. With the amount of antler that is produced in the United States today, if one percent of the population took one pill per day, it would run the antler supply totally out in ninety days. So even without selling meat, we're just scratching the surface of this industry." ▪

Lyle McLain has farmed for forty-seven years just four miles from where he was born south of Mohall. He has raised grain and livestock - cattle, hogs and sheep - and now elk. He is the president of the North Dakota Elk Growers Association.

The Reverend Dan Paulson
Trinity Lutheran Church ■ Alexander

Dan Paulson was ordained at age fifty after careers in the cabinet business, building and selling copying machine stands nationally, driving a beer truck, tending bar and selling spirits. Today he's still in the spirit business serving his first parishes in Alexander, Arnegard and Highland, a small rural church. Paulson's wife and spiritual partner, Kathy, is a native of Makoti. Both felt called to return to a rural area to serve.

"Fifteen years ago McKenzie County was one of the ten fastest growing counties in the nation. Fifteen years ago! Now the county is chasing to be number one the other way. But oil is showing signs of getting better again, and that does indeed give our farmers and ranchers, along with the oil workers, a breath of fresh air. We are very concerned for our children who have little, other than agriculture, to look forward to in McKenzie County. We need some help in bringing in opportunities for them. But the way it's going in agriculture our school systems are in dire trouble. We still hold out hope. The statistics say rural America is shrinking and will continue to do so. Yeah, that's true. I don't know that we're blind about that, I don't think we are. We have to trust. God has shown these communities all along that God's promise is true, that God does provide, always has and always will. That provision also means that we have to be innovative and change things from the way they used to be. Our farmers and ranchers aren't afraid of change.

"A few weeks ago I would have said the situation in rural communities like this one was somewhat hopeless. A couple of people have quit farming this year, but it's really the first time that has happened. Oil has always been the underlying helper and it's showing signs of coming back. But with farm prices staying low, some guys got out before they farmed away all their equity. Frustration is why producers want out. They have the ability to produce but not receive an equitable amount of return. It seems like the banks and government agencies want to put the brakes on farm programs, and the farmers who participate in the programs want out, too, but there's no place else to go. Farmers can't get out of debt by selling their farm, so they just go down deeper and deeper.

"If you know western North Dakota agriculture, you know a great amount of depression and despair. It's a tough way to make a living. But there's also a great amount of hope out here, too. In our congregation I see people who are determined, who trust beyond any human understanding. I think it's called faith. You know, the farmer gets accused of being the guy who always does it the same old way. I don't think that's true at all. Farmers have had to be innovative and have vision in order to produce as much and as well as they do. They have to know how to be good managers and good stewards and practice it. They're not afraid to take another job or two in order to stay on the farm. I find that really remarkable. Especially remarkable when you know how well they raise their children. They bring them to worship and Sunday school, and take the time to attend all their school activities. So many are trying so desperately to be good parents as well as faithful people of God. I see a lot of people doing it. They're working their tail ends off. I also know older people who should be retired and enjoying the benefits of working all their lives, and they are still hard at it because they have a big debt load to pay off.

"People here are strong spiritually and it helps them. No doubt about it. Part of what helps us is our connectedness as a community. We do a lot of things together. We don't necessarily like or love each other all the time. It's like any place else; we argue and cause each other pain. But every Saturday night in the summer we get together for hamburgers in the park, and its unbelievable. People talk to one another for hours about what's going on. I think these gatherings, and our church and the school, and the way of life people have chosen to live here, is the hope for the future. People here don't necessarily want to be rich, at least not monetarily, and often because of the pressure they're under they don't realize how spiritually rich they are. But they do seek to live this life because it spiritually fills them and gives them and their children and their community benefits that go way beyond financial gain.

"Another reason for hope here is because of the strong ministry that goes on in this county. We cross over denominational lines and have been really intentional in proclaiming the oneness of Christ Jesus in the lives of these people. God provides these people with a lot of fortitude and strength and when they come together as one, praising and praying together, with confidence in God's love and compassion, then new hope comes alive in them.

"I think the role of churches in rural areas is to be innovative. Maybe pastors will have to serve two or three congregations; many of us are already doing that. I drive more than 2,000 miles a month covering this area. I think as the community shrinks, the presence of the church as a constant in people's lives will be important. People can be assured of our want and desire and our will to follow God's call to serve them.

"I'm not a fixer. I'm a listener. All I can do is listen. God is equipped with the answers, and God brings them even if we don't like the answers. Individually there is nothing we can do to fix what's going on, much less turn it around. But together as one in Christ Jesus, we can listen and dialogue, and pray for understanding and vision. The future here will be stability but not growth. Bigger farms, sure. But maybe we can invest in some worthwhile opportunities to help our children at least have the option of staying here after they complete their education, because they don't want to leave." ■

With members of his congregation struggling through continued low prices for grain and beef, and the prolonged downturn in the oil economy, the Rev. Dan Paulson believes he must be deliberate in his community-building efforts. The strength of his congregation is their togetherness and ability to keep the lines of communication open.

John and Chris Skogen understand the risks in farming, but it's been their dream to one day take over the farm their grandfather started north of Epping. Both brothers work off the farm to support themselves until they can begin buying or renting land from their family. They believe they have the persistence to become successful farmers.

John and Chris Skogen
Optimistic Young Farmers ■ Epping

"We like the idea of being able to work on our own and make a living on our own. We want to be our own boss. Part of the attraction of farming is the freedom of running our own business and providing for ourselves.

"We think our farm is a good set up. We've been taught how to run things and how to plan. We use our heads and get the job done right the first time. We have a good opportunity to take over the farm. Being able to get in and go with the flow of things really helps. We hear stories sitting around the kitchen table about how tough things can be, but that helps us plan away from difficulties toward success. It helps to listen to a person who's been through it because in farming you have to be ready for almost anything.

"We don't think about bad prices. You just do what you can. You're definitely not going to make any money if you don't put anything in the ground. Prices have been low and they've come back up again. It runs in a cycle. There are ups and downs but you have to look at the averages. Farming is about riding it out. You just have to plan and ride it out.

"There are young people who would like to get into farming but they have no way to do it. They aren't fortunate like us to come from a farming family. The costs are just too high to start from scratch; it's virtually impossible to get in if your family isn't farming. Right now for us, our farm is a good size. Our family pretty much owns everything. Grandpa is seventy-six and dad is fifty-five and we are young guys just coming out of school. The four of us can operate a fairly good-sized farm. The amount of acreage is kind of relative. In the Red River Valley you don't need as much land to make it. Out west where we live the soil is lighter and we fallow half of it every year, so we need more land to live off. But, whether you like it not, a bigger farm is the thing of the future. Small farms just can't make things work anymore.

"We raise wheat and durum. It's worked in the past and it's put us where we are today. These crops are really optimal for our area. They grow well and we know that they work. Other guys are trying different crops, but down the road there should always be a need for wheat and durum.

"We're interested in farming because it's what we know, we like the lifestyle, and we're optimistic we can make it work. It's tough to get into any business with a pessimistic tone. If you're pessimistic about it chances are you're not going to make it. Our farm raised us, it raised our dad and grandfather and great-grandfather. We've come from a good situation that's provided well for us. We have good ideas and we know if we use everything we'll make it work.

"There's not many young people back home. Out of our class of thirteen in high school only one is farming. The others didn't even contemplate it; they knew a long time ago they weren't going to farm. They didn't want that lifestyle. We've been asked a lot of times, 'How are you going to bring a woman out there?' Hopefully we'll find someone who'll match up with us, someone who is going to work with us and have the same interests. We don't want to hook up with someone who would draw us away from farming, away from something we've always wanted to do. There are not a lot of young people in the area. There just aren't.

"We know what a struggle it can be. Farming is the biggest gamble there is. But you look ahead. The last two summers we've been breaking up land that's never been broke before. Spent two summers picking rocks. It looks like a pasture of sheep out there; there's so many rocks. But we picture that land with a great crop on it. We see it ripe with grain and the wind blowing. That's what keeps us going, knowing it'll be ours someday." ■

Recent graduates of Concordia College, Chris and John Skogen anticipate taking over the family farm in western North Dakota. Both single, the brothers will use their degrees in business and education to teach during the winter and farm in summer as they wait their turn to begin farming fulltime.

Norman Weckerly has been a leader in farm organizations for nearly fifty years, and has seen lobbying for change in Washington move from friendly visits with Congressmen to an adversarial relationship with expensive lobbyists and lawyers. In the meantime, enthusiasm to continue farming wanes in the face of prolonged poor prices and poor trade policies.

Norman Weckerly
Farmer and Businessman ■ Hurdsfield

"There's no question fundamental change is happening right now in agriculture. The thing is, it's affecting the assets of people so much that it changes lives and whole ways of living.

"I'm a proponent of getting government going on marketing. Get us competitive in the world or else buy us all out. Make a Buffalo Commons out of the area. Everything west of the Red River Valley — buy us all out. I've been involved in promoting agriculture ever since I started farming. For years, when a delegation of farmers went to Washington, like the Durum Growers or Wheat Commission, we were listened to, and in most cases our suggestions were carried out. Today, the only way you get listened to is by hiring a law firm. They lobby for you. No matter what is done in Washington today, it's adversarial.

"People in farming are wearing out. It gets hard to keep some optimism going out here when you're fighting the battle all the time. The thing about farming today is, when you plan to put in your crop you plan more to break even than to make a profit. You just plan not to lose any more money. You don't plan to buy equipment. You don't plan to expand or capitalize. You try to rent more land and use the same equipment you have. It's called 'risk management', the current buzzword. But by farming more land you just have the potential to lose more. Everybody thinks the answer is better crop insurance. There again, that just keeps you from not losing as much money. It's never a matter of making a profit.

"It's hard to keep your enthusiasm up when you know the best you'll do is break even. That's it. That's the problem today. You're always hoping maybe there will be a disaster someplace or something will happen so the market will go up. If you take all the risks in farming there should be the opportunity to make a decent return. Today there is no opportunity for that. If people want freedom to farm, then we have to have an opportunity to make a profit, otherwise we're going to need a guarantee that we'll make seven or nine percent every year and can't possibly lose. What's so ridiculous about this whole arena we work in is public policy is causing farmers to subsidize our cheap food.

"It's gotten so that people around here have had a hard time leasing out land. No one wants to farm it. Heavens, I never thought I'd see land go unwanted. The first twenty years I farmed, land is what drove us. Land was what our optimism was, because real estate values were increasing. Even though we didn't make a lot of cash money, our equity increased because real estate increased. From the time I started in 1956 up until 1980, real estate continued to increase. Since then it's either gone down or stabilized.

"In Wells County there are people throwing in the towel. The squeeze just continues. Earnings are so low that it's hard for some of us who have been optimists to try to encourage people that next year will be better. It's gotta get better, but it doesn't seem to be happening. We're not alone. It's all the commodities — oil, gas, timber, aluminum, you name it, and it's a world market. We had seven inches of rain here at harvest time, and it didn't make a blip in the market. That rain just increased our costs probably ten to fifteen cents a bushel. We lost production because of it. A lot of farmers in North Dakota are losing production because of this wet cycle we're in.

"With prices so low, you try to harvest more bushels. That means farming bigger pieces of land with big, expensive equipment. Then you pay more for inputs to control weeds and disease. So it's land, equipment and chemical. But there's no earnings. It's a cycle that doesn't end. You're in it and you can't get out of it. This is true for probably eighty percent of the operators in North Dakota. It isn't that they want to be out here, they just don't have any alternative. Who am I going to sell to? Who are we going to rent to? Nobody. I mean, who wants to invest in an agricultural operation today?

"The so-called experts say the problem with farmers is we don't know how to market our grain. That's ridiculous. A farmer today has to be an expert in many things. To run these farms we've got to be a mechanical expert, a chemical expert, a time management expert, a people management expert, a truck dispatcher and grain shipping expert, and a money management expert. If you're not good in just one of these areas, your operation suffers. All the applications we use — chemical tank mix, fertilizer blending — we're almost getting beyond our ability to adapt to all this technology.

"What about all this initiative farmers are supposed to have? Let me tell you, it takes money to carry out initiative. Personal initiative doesn't overcome a combine that costs $150,000 when your old one is worth $15,000 and it's worn out. It comes right back to the fact that any industry that doesn't have earnings has to restrict, retract or regroup, and that's what happening in all commodities. Agriculture is probably the worst because it affects so many people in so many places.

"I don't think there are any villains in this agricultural situation. I think it's big, structural change. It's a revolution in technology and how we apply technology. It takes earnings to apply technology and to advance. If you don't have earnings, you can't do it. There's no way to recapitalize. That's why there aren't any young people staying out here, and very few coming in. It's like one guy said, our town is just old people and broke farmers." ■

Norman Weckerly began farming in 1956. For many years he owned the local elevator that his son now manages in a family partnership. He operates the general store in Hurdsfield with his wife, Lou, selling groceries, farm supplies and hardware; and with his family, farms 10,000-acres. Weckerly is chairman of the National Bank of Harvey and has been active in a number of agricultural promotion groups.

Al Ulmer
County Extension Agent ▪ LaMoure

A native of Edgeley and a 1980 graduate of North Dakota State University, Al Ulmer has been a county agent for twenty years. He is past president of the North Dakota County Agents Association. County agents are an outreach extension of NDSU providing resources and expertise in the areas of agriculture, community development, and family and youth development.

"We need to be open to new ideas. When baby carrots first came out, people were saying, 'Who would want them? That's the stupidest thing I've ever seen.' But at our house we haven't bought a bag of regular-sized carrots in years. Baby carrots are more convenient, they're sweeter and our kids eat them right from the bag. That 's how farmers will have to start meeting new consumer demands. We need to imagine what our final product is going to be. Why are we still raising wheat, a crop we know we can't get rid of, and we're still complaining when we can't sell it? Ford isn't making Edsels any more, are they?

"Recently, economists have come out with a startling discovery that most of us already knew but were afraid to admit. For the last twenty-five years, after you take out the government support payments, wheat has been a 'dog' economically-speaking. The reason so many producers raise wheat is that Dad did it, and Granddad did it and Great-granddad did it. So I suppose that means the younger generation will raise wheat, too.

"There are some farmers who are willing to take risks and be the first to try something new. It's not age, necessarily, or education. Some farmers who've been to ag school aren't nearly as willing to try something new as some who haven't been to college. It's just some people are innately more eager to try something new. Some farmers have to watch a neighbor's new crop for three or fours years before they'll finally be willing to try it themselves. Farmers do watch what their neighbors are doing, and after they see something promising on the other side of the fence, they say, 'Jeez, it's working for him, maybe I should try it.' But then you also hear others say at the local coffee shop or bar, 'Yeah, so and so did that twenty years ago and it didn't work.'

"You would think it would be the farmer who is in trouble who'd be eager to try something new. But I've noticed they are often the ones who are inclined to stick with the things they're familiar with, which around here is wheat. They're probably familiar with wheat, have the right equipment for it, and it doesn't cost a lot to put in. If things work out, they might be able to keep ahead. Sometimes the higher costs of innovation will keep farmers from trying something new, but a lot of times it's just the comfort factor.

"When I find a farmer who's growing a different crop or using a different tillage method, I ask to have a field day on his farm. It's surprising, even though the number of farmers is becoming fewer and fewer every year, the number who show up for field days stays pretty stable. It's the same thirty-five to fifty guys every time. Some farmers just want to know how they can farm better.

"There have been some changes, though. In just the last six years in LaMoure County I have seen the number of wheat acres decrease and be replaced with soybeans and corn. Now we have about the same number of acres in sunflowers as in wheat, where wheat used to be king.

"There's another change coming. We're going to see more and more crops grown under specialty contracts. Those contracts specify what the farmer will produce, for example, a certain plumpness of the kernels, a certain amount of protein, certain germination standards and a certain style of irrigation so that the amount of moisture can be controlled. Contract farming will help farmers make money, but they're technically working for someone else.

"To survive, farmers will have to become more business-oriented. As one guy said to me, 'I have more fun when I'm out on my tractor, but I make more money when I'm sitting at the computer and doing my book work.' Farmers will have to make the management part of farming the most important part, instead of something they just have to do to keep track of things for taxes.

"But as we are becoming more business-oriented in order to survive and feed our families and stay living out here, I'm afraid sometimes there's one thing we might be losing - our sense of wonder at planting a seed and watching it sprout and come out of the ground. It's kind of like watching a miracle, like a calf being born, seeing it take its first breath, get on its legs and making sure the little bugger gets something to eat, gets up and gets going. We're in jeopardy of losing some of the things that makes farming special.

"In the next ten to fifteen years I think we are going to see two distinct types of farmers. The big farmers are going to get bigger. The big farmer, rather than driving a tractor, will be more of a CEO kind of farmer. He's going to spend most of his time at the computer following market trends, farm programs, things like that. What's more, he's going to hire somebody at a good wage to do the actual farming. The other kind of farmer will be small hobby-sized farmers, people who may have jobs in town but keep a few acres on the side. They'll grow specialty crops like vegetables, carrots or onions that don't require many acres. I think smaller farmers are going to find niches where they can fit in somehow.

"I think there's a transition we need to make if we're really going to understand what's going on in farming today. Instead of talking about the family farm and the farm family, we need to talk about families that farm or families who are in the farming business." ▪

One important function of country extension agents like Al Ulmer is to make available the research of the land grant universities like North Dakota State University to farmers so they can apply current research to their operations and reap the financial benefits.

It's a family affair twice a day in the milking parlor on the Scherr farm. The Scherrs milk seventy-five cows "and we're satisfied with that. It's about all the five of us can handle, but it's getting tougher every year. We can't figure out why prices are so low. We're out here living the life we want. If us small guys can't make it, who'll live out here?"

Tom and Joyce Scherr
Dairy Farmers ▪ Zeeland

"Our girls are always looking at their watches. They know when it hits three in the afternoon we have to be at home. It's time for chores. Our seven-year-old feeds the calves. We don't have to say anything to her. She's automatically out there when she's supposed to be, and she knows where to get the feed and what to do. Our girls know how to work. Whatever we do out here, it's all family labor. We make sure they know how important they are to this farm – we wouldn't be here if they weren't working with us.

"We've learned in dairy farming you don't do what you want to do; you do what the cows want you to do. The cow - she needs her grain, she needs her feed. When she walks out of the milking parlor, she wants fresh water to drink. How you treat the cow is how she's going to treat you. We feed top-quality alfalfa, so we'll get seventy or eighty pounds of milk per cow. That cow will want nice straw to lie down on. She doesn't want to be too hot or too cold. The thicker you bed her pen, the better she likes it.

"We have to be up to standard on our milking parlor. The walls have to be white. The milking machines have to be so clean you could eat your supper off them. There can't even be any holes in the floor that will collect water. That's how it has to be – completely clean and sanitary.

"What a lot of people in dairy can't handle are the hours. We milk at the same time every day of the year – at three in the afternoon and four in the morning. When you wake up in the morning, you can't push the snoozer and sleep a little longer. If you're a half-hour off, it's not so bad. If there's an hour or two-hour difference in milking time, we'll notice it right away. Production can drop maybe five pounds per cow. Our milk goes to a cheese plant at Pollock, South Dakota, about seventeen miles southwest of here. They make shredded mozzarella and cottage cheese.

"The first milk check we got – back in November, 1979 – was for $14.60 per hundredweight. It's been as high as $17. Right now it's just below $10. If the quality is high – low cell and bacteria counts and high protein – we can earn more. Our last check was close to $12 just because of our quality. One good thing the government has done is blocking the price of milk at $9.50. Milk can't go below that, but look at how costs have changed since 1979. Look at gas prices. Will gas triple while we still have these low milk prices?

"The low prices, the hours, you can see why people around here are quitting dairy. Some people got into dairy when prices were bringing around fifteen dollars per hundredweight. All of a sudden that drops to ten dollars a hundred. So a farmer goes out and starts looking at other jobs and finds he can get his insurance and retirement covered, too. He could go to work at eight and be done at five. He wouldn't have to worry about his paycheck taking a drastic drop. Well, he's going to quit dairy farming and take that job, isn't he?

"Out of this $10 per hundredweight we're getting paid, we have to come up with everything. We pay our own health insurance, we pay into our retirement plan and cover whatever our kids need. It all has to come out of this $10 per hundredweight milk. Our style of living has not gone up in about the last six years. Our health insurance has gone up, and food has gone up. Joyce works two and a half days a week at the clinic in Wishek, about thirty-five miles from here, as a receptionist and secretary. What she makes pays our health insurance and Alisha's car payment. When the girls aren't here any longer to help pull the load, I don't know that we'll do.

"People have said to us, 'Why don't you milk 500 cows? You could hire people.' But it's hard to find good hay or alfalfa around here. Besides, when an operation gets too large, like those big 3,000 cow outfits in California, people forget about quality. In a family operation, I think we care more about what we produce because we care about what we're doing as a family. When those guys get to milking 500 cows, do you think they're going to do it for $10 per hundredweight? People complain now when the price of milk goes up. With only large operations, milk will go higher because those huge farms aren't going to produce milk for nothing.

"Last year when the milk inspector was down from Bismarck even he wondered why people are keeping on with dairying and letting cows run their lives. But it's not that easy to just say, 'Oh, well, the price is low, we'll just get out. We don't care anymore.' We've been through droughts when you come home at night and just want to cry because you don't know where you're going to get feed. We've cried a lot together. But we've had a lot of happy times, too. We're finally getting the farm to where we wanted it to be twenty years ago.

"We're happy and we're healthy. We have good food on the table, a roof over our heads, and money coming. Gifts at Christmas aren't a big thing for our kids; they're more interested in who's coming to visit and looking forward to a big family gathering. In the summer, for us to go out and spend a Sunday afternoon together — that's good enough for us." ▪

Tom and Joyce Scherr and their daughters, Kristen, Alisha and Emily, were named top dairy family of McIntosh County in 1999 and were later chosen as one of six Dairy Family Ambassadors from North Dakota. They farm near the South Dakota border two miles south of Zeeland on land which Tom's father, Kasper Scherr, began dairy farming in the 1940s.

Michael McMullen
Plant Breeder ▪ Fargo

A professor of plant
sciences at North
Dakota State
University, Dr. Michael
McMullen has been on
the faculty for
twenty-four years.
McMullen's specialty is
oat breeding and
genetics, and his
research is focused on
improving oat varieties.
McMullen also serves
as a consultant to
farmers exploring
ways grain research
can help them improve
crop quality.

"Several years ago when U.S. farmers were getting a dollar and a half per bushel for oats, Swedish farmers were getting about four dollars, with the difference made up by government subsidies. That sums up the view Europeans have of food production. To Europeans, self-sufficiency is very important. They would rather raise their own food even when it would be cheaper to buy from another source. It goes back to times of crisis, like World War II, when Europeans experienced serious food shortages.

"Ag policy in Europe today means that farmers are making a living on 200-acres and doing well. It also means their people are supporting farmers who wouldn't survive without a subsidy. Apparently they believe it's worth the price.

"Plant breeders sometime get blamed for the problems of overproduction. By working for higher yields we have helped create surplus. But if American plant scientists hadn't bred for efficiency in crop production, we can be sure the Canadians or Europeans would have. Besides, our research has resulted not only in better yields but also in higher quality grain.

"Quality is the thrust of our research. We hope quality will bring buyers to North Dakota rather than to Canada or the Scandinavian countries. Right now a U.S. farmer is paid for test weight and hull percentage but not nutritional quality. Oddly enough, when it comes to nutrition, Americans pay a lot more attention to what we feed our livestock than what we feed ourselves. Even though farmers aren't being paid for nutritional quality for humans, plant breeders feel a moral obligation to work to improve it. As buyers become more aware of nutritional factors, we hope nutrition will become an important factor in selling grain.

"Maintaining quality also means we spend a lot of time developing varieties with disease resistance. The pathogens that attack grain keep changing, that's why it's a constant battle to stay ahead of them. Sometimes even less than a year after a variety has been introduced, we find it being attacked by a pathogen – nature has a way of finding any niche that's available. We hear complaints that there are too many grain varieties to choose from, but there needs to be several varieties available in order to choose the best-adapted ones and to avoid genetic vulnerability. If mostly the same variety was planted over a wide area, and it turned out there was a genetic weakness rendering it vulnerable, it could mean a large-scale crop disaster.

"The Scandinavians have a point with their high government subsidies. I think farmers will always need a safety net. Just-in-time supply – goods coming to consumers just when they're needed – is efficient and works well in some parts of the economy, but not in crop production. I'm not sure I want to rely on what was harvested yesterday for what's going to be on my table today. If production should lag by just a day, we would be hungry. In the mid-1990s we had some of the lowest food reserves in history, with at times only a sixty-day supply of grain. We need a certain amount of overproduction for security, yet overproduction leads to depressed grain prices. But overproduction is not nearly as much of a problem as shortages, so I see government price supports as insurance.

"The outlook for grain prices is bleak for the next five years, but I think the time will come when the U.S. is asked to produce more. A great deal depends on a general improvement of the economies around the world, particularly Asia. The Chinese have a goal of increasing poultry production to provide one egg per capita per day as a protein source. To achieve that will require an increase in grain production equivalent to all the grain now produced in the Prairie Provinces of Canada. Economists who suggest the United States can drastically decrease agricultural production and import most of our food are possibly right in the short term – but probably wrong in the long run. Again, that's why we need strong government support – to keep our production infrastructure in place so that when the world needs food, we'll be able to provide it.

"The worry of widespread crop failures is always lurking out there. Some of us find it alarming that in our own time, while total production of grain has increased over the past few decades, per capita world grain production peaked in 1984 and has been dropping since. Hunger is occurring not because we can't produce enough food but largely because poor economies keep some countries from being able to buy the grain their people need. We have been fortunate over the past several decades to have good worldwide production without many major crop failures, but that could change.

"So far, genetically modified varieties have improved weed control and insect resistance, but they have not put more money in farmers' pockets. Many farmers who have tried genetically modified seed haven't felt that they were any farther ahead. We're seeing decreased genetically modified production, thanks in part to consumer reaction in Europe. In the long run we can be pretty sure genetically modified crops will be accepted, especially when the time comes when we need all the production we can get to feed people." ■

Dr. Mike McMullen examines kernels of oats that are nearly ready for harvesting on test plots at NDSU. Patience is a virtue for researchers like McMullen, who has worked on this disease-resistant variety for nine years. It is still another two years from commercial release.

Robert Carlson
North Dakota Farmers Union ▪ Glenburn

Elected president of the largest farm organization in the state in 1997, Robert and Mary Carlson raise bison and small grains. Carlson has been active in many agricultural organizations, including the National Farmers Union, USDA Agricultural Trade Advisory Committee for Grains and Oilseeds, United Spring Wheat Processors, Dakota Growers Pasta Cooperative and the North American Bison Cooperative. Much of his time is spent working with government officials to secure federal action for North Dakota farmers.

"Freedom to Farm was a lot like the Titanic. No one seemed to think anything could go wrong with it. When we enacted it in 1996, many people in agriculture bought the notion that weaning farmers from government support – or "government interference" as some put it – was the wave of the future. Increased global trade would take the place of government safety nets. We were entering a new age of prosperity. Some farm groups found themselves fighting for a deck chair on the Titanic. The iceberg in this case was low prices.

"Freedom to Farm was passed during a time of good prices. Producers tend to be optimists; when prices are high, our memory of what happens doesn't seem very long. Whenever we get a period of high prices in agriculture, the immediate response is to produce more. Then the price plunges. So we soar high, we stall, and we crash. That's what has happened.

"Suppose we continue on the path that was charted in 1996, and we reduce barriers to agricultural trade and allow for the free movement of agricultural products around the world. Suppose, too, that at the same time we reduce government support for domestic agriculture. What that will mean is that the low-cost producers in the world will be providing the raw agriculture products for the processors. It will be a race to the bottom.

"If you carry out this notion that free markets and free trade will be good for the United States and the world, you're really playing a dangerous game with food security. If we would have free trade in toasters, for example, and some day there's a crisis and there aren't any toasters for sale in the stores, that's an inconvenience. What if, however, over time the free market goes to work and by a process of elimination causes several events to happen: land continues to go out of production, farmers go out of business, supply becomes roughly equal to demand, and then there is a natural disaster somewhere in the productive part of the world. In this picture we could have a real food shortage and a real crisis on our hands. Those with the most money would be supplied, and those who couldn't afford it would go hungry. I've wondered what would happen, if for just one day you went to the grocery store to buy pasta and there wasn't any on the shelves. Would you think about agriculture in a different way? I'll bet you would!

"The assumption among the free-trade theorists, and those behind Freedom to Farm, is that when prices get low, demand should increase. Surpluses will be eaten up somewhere, and with a harmonious relationship again between supply and demand, prices will rise. That's the classic economic model. The flaw in that argument is that food is not like other commodities. If bread, for example, was half price tomorrow, you might buy a large quantity of it and put it in your freezer. But chances are you and your family would not double your consumption of bread just because it's cheap. Food is too important to be left entirely to free-market forces. Europeans have recognized this for a long time. They, of course, can remember being hungry in two world wars.

"I think we will recognize this race to the bottom, this race to become the cheapest producer, won't be won by the United States. That's because our costs are higher, and we demand a certain standard of living that other people in the world might not demand or expect. Farmers want to be able to send their kids to college and have a decent quality of life, too. If we enter a global competition to see who can be the low-cost producer, it won't matter how large your farm is; in the United States, you just won't be able to do it.

"What we need to do in our trade negotiations is shift from removing support mechanisms from agriculture to tying support mechanism to some sort of an arrangement to bring supply and demand into harmony. For that, we need to do two things. First, we need to establish an international food reserve to draw from in times of production disaster. Second, we need to engage in some sort of set-aside program on a world basis. If stock carryovers are burdensome, exporting nations can reduce their plantings and reduce their supplies to keep supply and demand in rough equilibrium, and get a market price that can enable producers to make a living.

"One of the opportunities for the future of agriculture in North Dakota is that people around the world are becoming more sophisticated in what they want in food. There are all kinds of niche markets developing. We need to link with the buyers and supply them with exactly what they want. The members of the new Dakota Pride Cooperative are raising special varieties of wheat for the State Mill, which they're paying a premium for. The mill can put the right wheat varieties into their flour blends to provide customers with exactly the kind of mixing and baking characteristics they want. Farmers are becoming much more aware that to stabilize their operations, they're going to have to get more into the retail and processing side of the business. Simply producing a raw product is not going to allow us to prosper in these markets. In the future, I see us growing a much greater variety of crops than we grow now." ▪

From his office at the Farmers Union headquarters in Jamestown, Robert Carlson keeps a sharp eye on farm developments, particularly in Washington. Carlson believes placing farmers into free markets and world trade may become an unnecessary risk for United States food consumers.

Deb Lundgren
Farmer and Legislator ▪ Kulm

Because of her concern for the worsening farm crisis, Deb Lundgren joined the Farmers Union and Democratic Party and in 1998 was elected to the state legislature. Lundgren grew up in Washington, Oregon and Michigan but became involved in farming after she married Blaine Lundgren, a Kulm native. In addition to the small grain farm they and their four children operate, she is a volunteer paramedic and runs a travel agency on Kulm's Main Street.

"I used to be so afraid to speak in public. In high school I tried not to get straight A's so I wouldn't have to give the commencement address. But in April, 1995, when Secretary of Agriculture Dan Glickman was coming to Bismarck to discuss the farm crisis, someone from the Farmers Union asked me if I would make a few comments as a farm wife, and I said I would. It was a very short presentation and I was shaking as I spoke. When I finished, the audience gave me a standing ovation and Congressman Earl Pomeroy asked if he could have my comments inserted into the Congressional Record. I was overwhelmed.

"Several months later we were at home having dinner and the phone rang. I picked it up and a voice said, 'Hi Deb, this is Byron Dorgan. I'm having a meeting next week and would like you to come and make the same comments you made to Secretary Glickman. Would you be willing to do that?' I told him I'd be happy to and asked where he would meet me. 'At the White House,' he said. 'We're going to meet with President Clinton and discuss farming.'

"I was amazed at how well President Clinton comprehended farming. Some people in public office have no real connection to farming, but he seemed to know what our problems were. I just sat there in total awe that a farm wife from Kulm, North Dakota, could be in the Oval Office describing the farm crisis to the President, but he made me feel very much at ease and was very interested in what I had to say.

"What I told Secretary Glickman and the President was that I enjoyed being a farm wife but that the farm doesn't support us anymore. Instead, we are supporting the farm. At one time that same farm supported my husband's grandfather and his nine children, and they lived there very comfortably. We dug out his grandfather's tax return from 1957 and compared it with ours from 1997. Would you believe the difference in income for both years was less than a thousand dollars? Forty year's difference and it's still the same income. What's wrong with this picture?

"Blaine started farming in the 1970s with his father. We did okay during the 1970s and 1980s. But toward the beginning of the 1990s we had water on the land for several years and prices had been depressed for so long that we wondered if we should keep on farming. There was no alternative than to have one of us working off the farm. At the time I talked to Secretary Glickman and President Clinton, I was working three off-farm jobs and Blaine was working two.

"President Clinton told me that we can do right by our farmers if we can keep hanging tough and if I'd keep speaking out. That has been pretty hard to ignore. I know there are people here with sixty years of farming experience who are devastated – not just down in the mouth, but clinically depressed. No matter how good they get at farming or how many technological improvements they take on, they can't make it. Do you know a majority of people in this area between the ages of thirty and fifty have no health insurance? It's no wonder we don't know of a single farmer that is encouraging his kids to go into farming. When I talk to people and they know I'm a legislator, many feel a little freer to share their fears and tell me to please try doing something about the problems in agriculture. People from both parties have been so gracious about thanking me for taking on the issue and for speaking about what is really going on and for trying to do something about it. And I'm doing everything I can.

"There are those in government who believe strongly in state and local control, yet on issues of agriculture, they say there's nothing the state can do, it's a federal issue. But right here in North Dakota we can fashion policies that will help farmers, like the wheat pool proposed in the last legislative session. If we as a people agree that a problem is a high priority, we'll find the money and the solution. We always do. Go back in history and you'll see how forward thinking our state can be – for example the Nonpartisan League. Farmers were told by the powers that be in Bismarck to go home and slop their hogs. The farmers got angry and decided to stand up and do something about it. That's how we got our state elevator and state bank.

"Politicians like Republican Senator Richard Luger from Indiana say that if other people have had to get out of professions they liked and are good at, why can't farmers? With farming conditions so bad, why don't we find jobs elsewhere? But what are Americans going to do about food? Do we want to surrender it all to those who have money or to the big corporations who will run the farms with a bunch of employees? Look at what these big corporations have done in the past. They start in one state then move on to another, and in their wake they leave a mess – sterile soil from concentrated use and a decaying infrastructure.

"Are we really willing to depend on other countries for our food? I would rather be a self-sufficient, independent country than say, 'Well, we just see no way to support the family farm, so we're just going to be at the mercy of other countries.' I'd like our government to stand up and say we need to grow our food in this country. The farmers I know are doing absolutely everything they can. They are already the best at what they do, or they wouldn't still be here." ▪

Deb and Blaine Lundgren stand amid confection sunflowers they are growing on their farm near Kulm. Since being elected to the legislature, Deb has increasingly been asked by her constituents about state help for farmers. People ask her why are agricultural issues being overlooked when farming is the state's major industry?

Scott Steffes chants the sound that sells during a large consignment auction at the West Fargo Fairgrounds.
Steffes says these sales help farmers acquire the specific equipment they need for their farms while at the same time they
can sell off implements that are no longer needed.

Scott Steffes
Auctioneer ▪ Fargo

"Back in 1960, my father's first auction sale was five miles from where he grew up. They sold the pigs, chickens, farm machinery and silage in the pit and grossed $5,500, about average for a fair-sized auction in those days. Today our average farm auction is in excess of $200,000. We travel farther to do fewer auctions but for more money. There are just fewer and fewer farms and people on those farms.

"From 1979 to 1986 is the period I would call the true farm crisis. The difference between what is happening today and what happened in the 1980s is that back then most farm families who were having financial troubles felt a strong attachment to their farms and wanted to stay on the land. But a number of them were forced to leave. So many people we were working with then did not want to sell out. They had a lot of animosity toward bankers, machinery dealers and elevator companies for refusing to give them credit or make them more loans.

"Since the early 1990s we have been seeing people leave their farms not because somebody is telling them to, but because they want to. They usually have a better understanding of the situation and just don't see much opportunity in staying on the farm. And when it comes to selling out, they are less emotional and more analytical about it. As they plan to sell out, their big question is how can they maximize their profits.

"It's amazing how many sales there have been in the last several years where we've sold lots of farm machinery but the owners are keeping the land. I'd say we auction off the land with the equipment and everything only about five percent of the time. People know that the roots of agriculture and their ties to their family farm are with the land. We're also seeing a number of people with rural roots coming back and investing in rural real estate. Another thing that amazes me is the age of buyers of agricultural real estate – people who are sixty or seventy years old are buying farmland. Sometimes it's because they're afraid of the stock market. Sometimes it's for taxes, but it's helping hold up rural real estate values.

"Ninety percent of the work of an auction is done before the sale. One of our biggest jobs in taking people through the process is encouraging them and counseling them, helping them see that there is life after farming and it can be wonderful. There are so many people who have put their heart and soul into farming that they don't see anything else. When you put yourself so deeply into your work, you become attached to things, like machinery, and even though it's made of iron and steel, it can take on a personality. I can't tell you how many combines I've sold for farm couples where the wife has put in thousands of hours on that machine and had an emotional attachment to it. With live-stock it's even worse because we are dealing with living, breathing beings. One reason why my father has been so successful as an auctioneer is that he understands what people are going through and can relate to them.

"The cycles of our business begins with doing an auction for a farm family when they retire from farming. A few years later we're doing another auction with them when they no longer want to take care of their thirty-acre farmstead or hobby farm. The third stage is when the farmer passes away and the children come back from all around the country to sell off what is left. Another cycle is what I call successorship. Let's say a successful farmer has a promising son who would like to take over the farm. We might do an appraisal for their operation so that the wealth can be split fairly among all the children. Pieces of machinery that the son isn't going to use might be sold. For example, the father might have had two 250-horsepower four-wheel drive tractors and two smaller combines. The son might decide to sell those so he can buy one 400-horsepower four-wheel drive tractor and one brand new John Deere 9610 combine that will do the work of two. They may opt to sell what they don't want at a consignment sale. But we don't see as many of these sales as you might think — not that many children are interested in taking over their parents' farm these days.

"Machinery consignment sales is another example of the evolution in farming. We hold them three times a year at the West Fargo Fairgrounds. In today's farm economy, because of the size of the farms and the special needs of certain crops, farmers constantly need to get rid of and acquire machinery. It used to be an unneeded implement might be left to sit out in the trees, but with the price of machinery today, a farmer can't afford to do that. With consignment sales, instead of working for one farmer, we're working with as many as 120 farmers who bring their machinery together for us to sell. It's working very well.

"Nothing stays the same, in farming, the auction business or anything else. Some people embrace change and react accordingly while others hold on to the status quo. There is a lot of negative thinking today, but I like to think in terms of how we can make agriculture better. There is nobody in the world that is better at what they do than our farmers. If you just allow them the opportunity – and that's all farmers want is an opportunity – look out, because great things are going to happen." ▪

Scott Steffes grew up on a farm near Arthur and went into the auction business with his father, Robert Steffes, who was elected to the National Auctioneer's Association Hall of Fame in 1999. Scott now has more than twenty years of farm auction experience, and regularly conducts workshops throughout the United States on farm appraisal and auctioneering. In 1995 he won the International Auctioneer Championship.

Roger Synek began studying to become a veterinarian but returned to operate the family farm after the death of his father. Successive bad years helped him make the decision to get out of farming. His experiences on the farm help him relate to his parishioners who are struggling with continued poor prices and high expenses.

The Reverend Roger A. Synek
Former Farmer ▪ Center

"I've learned that peace and serenity in the middle of turmoil is God. I remember sitting on the tractor agonizing over the question of selling out. I prayed the Lord's Prayer very, very slowly, and that forty-acre field went by so quickly. It was helpful to me because it gave me rest from the pain of figuring out what I was going to do and how I was going to make this decision. I discovered praying can give us peace of mind by doing what we can and including the Lord as part of the process.

"Farmers spend a lot of time on their tractors, and that can be a good time to meditate and pray. Praying can be just talking to God in your own words. People can praise God for all they have and trust that He will take care of them. I used to notice seagulls flying around my fields and think about how God made that big wing span so they could soar. If we're not obsessed about how to keep things going, fieldwork can be a time to let our minds drift to God. I believe people are more likely to find God in nature than in a computer screen.

"The Catholic Church has taken a strong stand on supporting the family farm. It's not that we believe farmers and ranchers are necessarily superior to city people or that there is something about just being out in the country that is good in itself. We believe farmers are carrying out a special kind of stewardship because they are giving of themselves to provide the rest of us with what we need to live – our daily bread. Their vocation is one we believe is especially blessed by God. That's why we're concerned about agribusinesses whose profits are excessive and at the expense of the people who actually work the land; people who are rapidly becoming part of America's economic poor. We support social and economic policies that include just compensation for those who work the land.

"The family farm does not guarantee that a family living on that land is going to work well together, but farms and ranches are one of the best guarantees of a healthy community. You see how that works when a parent takes a son or daughter along on the tractor. Riding together like that strengthens the family. Children are learning what their parents do when they're out in the field and they're bonding. Children know they're not alone and they're learning trust. And parents have the satisfaction and can praise all that they have when they look over and see their children beside them in that tractor.

"A farm or ranch is a good place for children because they learn responsibility. They have chores to do. We lived eighteen miles from town, so even though I was driving at an early age, getting into town for sports or other activities wasn't easy. The farm taught us, no, we can't just go into town when we want to, or that we had work to do for our livelihood, so we had to stay home. Farm people miss out on some things, but we get to work together as a family.

"I'm concerned about children today who see their parents stressed not only at harvest time, but all year round. Older children probably have learned to cope with it, but I wonder about the little ones. And so many farmers and their spouses must take off-farm jobs to try to stay afloat. I don't think all that time apart is good for family life.

"I found farming takes a great deal of careful thought and planning. There are so many questions and so many unknowns. But I think the more we try to control things, the less control we actually have. It's all that obsessing about the weather and things beyond our control that gives us mental pain. That's what I found myself doing, especially about when to sell my crop. Farmers have to learn to look at all the factors, read what signs there are, and pray for guidance; and then realize they've done the best they can. In Proverbs we learn that there is time for everything. We can also say there is a time to think about what we are going to do, and there is a time when we should stop thinking about it, too. The biggest thing for us to remember is God will provide for us, always. It may not be in farming, but He will provide. The skills we learn in farming will carry us for the rest of our lives. By keeping God in our lives there will be serenity and peace.

"I was two years into the pre-veterinary program at North Dakota State University in 1982 when my father died. I decided to take over the farm – about one thousand acres of cropland and a hundred head of beef cattle. For awhile my brother helped until we decided there was no way we could both make it on that size farm. To make ends meet, I worked three off-farm jobs. I swept up metal chips at a machine shop, pitched manure at a horse farm and worked at the sales ring. Usually it wasn't until everything else got done that I could get to feed my own cows, so I seldom saw them in daylight, especially in winter. I was getting pretty depressed.

"The last straw came in 1987. It looked like a good year, but at harvest it started raining and it rained for six weeks straight. The wheat sprouted in the field before I could get to it. It was mid-September before I got my grain in the bins and then the tractor motor seized up. That was it. I had tried so hard to figure out how to make ends meet, but it just wouldn't pencil out. I finally said 'enough' and scheduled the auction. It was only with God that I was able to get through all that pain. I sensed God had been working with me those five years on the farm while I struggled with what to do with my life. I went back to college and decided to answer His call to become a priest." ▪

Father Roger Synek grew up on a farm near Williston and eventually farmed his home place. Poor conditions in the 1980s caused him to get out of farming. In 1998 he was ordained a Catholic priest and he now serves as pastor of St. Martin's Church at Center and St. Edwin's Church at Washburn.

Albert Two Bears
Rancher ▪ Cannon Ball

Albert Two Bears raises cattle and horses along the shores of Lake Oahe. He served for sixteen years on the Standing Rock Tribal Council, working to construct a rural water system for Cannon Ball, provide HUD housing and build the Prairie Knights Casino. His wife, Sharon, works at Standing Rock Community College. They have four daughters and one son: Karol, Jaci, Allyson, Donna and Don; and four grandchildren, D.J., Cody, Jaye Don and Lindsey.

"I've been a cowboy all my life. All I ever wanted to do was be a rancher. When I was growing up in the mid-fifties we had cattle, but things got so tough dad had to sell out not long before he died. I can remember driving our last cattle down to Cannon Ball and shipping them out on the railroad. We filled up two stockcars, shipping the cattle all the way to Sioux City, Iowa, to sell them.

"A neighbor south of here, Aaron Gullickson, taught me ranching. I worked for him when I was in grade school and high school, hauling bales, working cattle and putting up hay. That's how I bought clothes for school and had a little spending money. With dad gone, it was mother who supported me during those years.

"I knew there had to be a way for me to have a ranch some day. Sure enough, when I was ready to start ranching, the government agreed we hadn't been justly compensated for the land they took for the Oahe Reservoir. In the early 1960s our tribe received more money, and one thing the tribal leaders did was set up a cattle program. I got thirty cows and one bull. We could pocket eighty percent of the calf crop, and we had to give back twenty percent until we had enough heifers to actually get going on our own. Then, after a few years, the tribe developed a revolving credit program where they would lend money for us to buy cows. So I borrowed money and bought cows.

"I've been a rancher ever since 1962. Today I have 450 head of beef cattle. Whether I lease my land or own it, I treat the land as if it's my own. You take care of the land, and it will take care of you. Whatever I do, I do with the sense that there's always a tomorrow, and we're going to have to use this land. It's not only me, but the generations to come. That means rotating my cattle to different pastures and partitioning off parts of my land so we don't just use it up. All those chemicals the farmers are using nowadays – do we know enough about them? Do we know what they are really doing to the land? It scares me.

"This is a bad time to get into ranching. There's no credit available for young people, as there was for me in the sixties. Since I'm an established operator, Norwest Bank at Mandan has helped me out quite a bit, but banks don't want to lend money to new operators. In ranching it takes years – at least ten or fifteen – before you realize any dollars. When you start out you have to spend money on equipment and you have to pay for your land. The biggest costs are equipment and livestock. Just to make it through those first ten years, somebody in the family will need to work hard at a job in town or somewhere else. When I first started, I drove heavy equipment for the government. My wife stood by me through all the tough times. She's worked just about all the while I've been in this ranching business.

"I don't know of any young folks today who can wait that long for a paycheck. A lot of young people have seen the hard times some of us have been through and they just flat out say, 'Hey, I don't want to put myself through that when I can get a good job in town, work eight to five and have weekends off.' I can't even name one young guy from among us Indians from Cannon Ball who is out on a farm or ranch today. Unless the tribe steps in and offers a lending program, the future of farming or ranching on the reservation is pretty dim. We could use some money like the Oahe settlement to help beginning ranchers and small businesses. It just so happens there's a big lawsuit against FHA right now for discriminating against Indians, so who knows?

"With this bad ranching economy we're losing the Sioux way of life. There was once a lot of Indian cowboys who rode the plains. We called them Prairie Knights. That's also the name we chose for our casino. Without Indians being able to have horses and some livestock, that lifestyle will be lost. Typically, we Indians on Standing Rock are not farmers, we're ranchers. We put in a little feed grain and plant some alfalfa for our own use, but we don't plant wheat, sunflowers or corn. When cattle prices stayed low about ten years ago, a lot of Indian ranchers I know, or their wives, took jobs in town instead of diversifying their ranches.

"Our family always had horses, so it was natural for me to go into breeding horses and raising bucking stock to sell to rodeo contractors. I started with quarter horse mares, and then I bought a really big blue roan stud, a bucking horse. I'm hoping to get some size into them by crossing a large Percheron with the mares from the blue roan. I've got a couple of bucking chutes on my place, so right from when they are young I start working on the horses to stay calm in the chute. I've had horses at the Mandan Rodeo, which is the biggest in North Dakota. My goal is to take a load of horses out to Las Vegas or Miles City some day for those big bucking horse sales.

"I'm amazed at how many changes have taken place in the short time since I was born. Back in the 1940s when I was a child, we didn't have electricity, indoor plumbing or running water. We got our drinking water right from the Missouri River. We'd just back our wagon into the river, fill up our barrels, and bring them home. We used to do everything with a team of horses and a wagon. In my lifetime we've gone from those days to electricity, color television and computers." ▪

Albert Two Bears loves horses and the ranching lifestyle. He runs a nice cow-calf operation but is most proud of his efforts at breeding bucking horses. He works with his horses so they stay calm at rodeos, yet know they are expected to explode from the chutes when the gate is opened.

Gary Hanson and one of the big, four-wheel drive tractors he sells to farmers who prefer the powerful machines for working the heavy soil in the Red River Valley. Hanson says implement dealers need to virtually become partners with farmers so they can properly maintain the high technology equipment in use today.

Gary Hanson
Implement Dealer ▪ Grafton

"Who would have imagined what was coming in farm machinery – how much larger and how much more sophisticated it would be. When I was growing up we never thought satellites out in space would guide planting, fertilizing and harvesting. Now there are devices on combines that tell farmers what their yields are on the go. In a few years we might be marketing a driverless tractor!

"All this new technology as well as larger farm machinery has made it much more expensive to farm. Farmers have continually asked for more technology. On tractors it's been smoother-shifting transmissions, quieter cabs and more settings that can be made from the cab. Larger farms need bigger equipment. I'm not sure farmers realized how this would affect prices. A combine that cost $50,000 in 1980 has almost tripled in cost, to around $150,000 today. If farm prices were better, it wouldn't be such a problem. But we're working with 1950s grain prices and the costs of twenty-first century technology. Farmers are running their equipment longer, and I don't see people trading as often.

"Technology is only one way the farm implement business has changed. Today farmers are savvier about machinery, more inclined to research and study the equipment that's available. Many of them use the Internet. The brochures from Case-IH now contain more detailed information and are clearly geared to smart customers. With both farming and farm implements becoming more complex, dealers have to know each customer's farming operation well, in fact, almost become a partner in each customer's operation. Farming in our region is much more diversified than it used to be, especially with alternative crops and new cropping methods. For farmers, diversifying can mean buying new equipment, or it can mean adapting what the farmer already owns.

"It was different decades ago, when nearly all farmers raised pretty much the same crops and we could have a large stock of just a few items available. Inventory is a big expense in our business. What we carry now is a broad selection of many items but fewer of each item in stock. We encourage our customers to plan ahead for buying equipment so we can place the order and have what they want available when they need it. The more lead time the better. We offer discounts for farmers who order ahead. This has been tough for many of our customers to adapt to, especially those who were used to the days when we would carry a lot of equipment in stock and they could just walk in and ask for what they wanted and generally get it in a day or two.

"Maintenance is another area where farmers need to plan ahead. With operations so much larger and equipment much more sophisticated and expensive, breakdowns can be costly. We just can't field a big enough staff in summer to handle routine maintenance calls. We urge our customers to bring in their equipment for scheduled maintenance during the winter months. It's just sitting in the machine shed anyway. We offer specials and discounts for overhauls, updates and maintenance of all kinds to get the farmers in. When it comes to equipment, everything needs to be planned ahead more than it used to be. We also have our repair people on call after hours and weekends during the busy seasons. We've always done this, but it's so important with today's sophisticated equipment. Our service people have to know the technology of modern farm machinery with all the computers and sensors. Our training is nearly continuous.

"The farmers who have suffered the most from this price situation are the young ones who are trying to expand. They can't afford new equipment, and because more farmers are turning to used equipment, that's getting expensive, too. Prices for grain haven't changed for years. That's why a lot of younger farmers have gone to hiring work done, hiring custom cutters and custom applicators, until they can afford to buy their own machinery.

"What has all this done to the implement business? We're losing dealers just as we're losing farmers. It all ties together. In 1973 there were 340 dealers in North Dakota; now there are 140. Dealers who didn't want to adapt to change ran many of the businesses that are no longer operating. In the mid-1980s many didn't want to invest in computers. Back then we paid around $100,000 for a state-of-the-art computer system, and then after five or six years we needed to upgrade. That's expensive. Dealers are also being asked to invest in buildings and new technology. These are difficult decisions to make during a down cycle in farming. Sales volume is another issue. Manufacturers set minimum amounts of equipment that must be sold each year for a dealership to continue carrying a line and being able to survive in the market.

"We just bought the Cavalier dealership. The owners weren't getting the return they wanted on their investment and decided to close. Case-IH asked us if we'd take over the operation. We see it as a chance to expand our base. We used to consider our territory as running from twenty to thirty miles from Grafton. Now our territory runs 70 to 100 miles in all directions. The Cavalier operation is a sublocation of our Grafton dealership, and it includes a service crew just like at Grafton. It makes good sense economically for us to spread out management costs over more locations. There are a number of advantages that come with size, including better training for our staff and maintenance facilities. What we're doing represents a trend in the industry. There are fewer dealers but a large number of sublocations out in the country." ▪

Gary Hanson grew up in the McVille-Kloten area, and after graduating from North Dakota State University he worked for Farmers Home Administration and Norwest Bank at Grafton. He took over the International Harvester dealership after it filed for bankruptcy in the early 1970s. In 1999 he was elected president of the North Dakota Implement Dealers Association.

John Fiedler
Former Farmer, Banker ▪ Mott

John Fiedler took over the family farm after graduating from North Dakota State University in 1979. He also worked at Farmers Home Administration where he gained insights and perspectives on the local economy. He left farming when Conservation Reserve Program payments became higher than cash rent. He is now vice-president and ag loan officer at the Commercial Bank of Mott.

"The years I farmed with my dad were great. We were raising small grains with some specialty crops for rotation purposes. We were holding our own until the CRP program came out. We were paying twenty-five dollars an acre cash rent to various relatives, but as CRP developed, our landlords realized they could get more from CRP than we were paying in rent. At that time, many CRP contracts were accepted at thirty-five dollars an ace on a ten-year contract. A lot of people went for it. Enrollment in CRP was substantial and today I'd say Hettinger County has approximately 100,000-acres enrolled.

"Once our relatives enrolled in CRP, our operation went from thirty-five hundred acres to around sixteen hundred acres. We analyzed our situation and tried our best to forecast into the future. The numbers didn't look good. I was concerned my parents would not be able to enjoy the comfortable retirement they deserved. There had always been generations of Fiedlers on this farm, but we had to be practical. With limited acres left to farm and rented land going for a premium, dad and I made the decision to sell our machinery. On the day of the auction I realized it was going to be the longest day of our lives. There's an emotional attachment to equipment and assets that comes from being in farming for fifty years. Dad wanted to keep the farm going, but I realized it would be difficult to buy his equipment, rent land and maintain a reasonable standard of living for my family. I had ten years experience at Farmers Home Administration, mainly working winter and spring months, so I decided to pursue a career in ag lending.

"The hardest part of my job is working with people I've known all my life. Telling a long-time friend his farm is not cash flowing is one the most difficult things you can do. It's not easy telling a friend he's lost net worth and it will be difficult to continue financing their operation. Agriculture is so information dependent and with these marginal economic conditions today it's been difficult for many operators to maintain profitability. Our bank uses government programs, like the FSA Guaranteed Loan Program, to help our farmers remain viable.

"Farming used to be a way of life, but now it's a big business. Some of our producers have decided to partially liquidate their operations. They'll rent out their cropland, keep the livestock portion of their operation, and maybe try find an off farm job. There's still a part of agriculture and being on the land left in their lives and they keep their pride. The land is rented to another producer who is getting bigger and spreading his costs over additional acres.

"In the last two years I've worked with fourteen farmers who were getting out or partially getting out of farming. That's an unusually high number, but we've had some bad years here. We had low commodity prices and poor quality crops in the western part of the county. Some guys didn't have the proper insurance coverage and it turned into a nightmare for them. People were getting seventy dollars an acre return on more than $100 an acre invested. If the government hadn't been paying loan deficiency payments, a lot more operators would be going out of business. It's been tough.

"People talk about what CRP has done to small towns. We used to have three implement dealers, a tire shop, welding business and a bulk fuel dealer. These businesses are gone and we now have to drive farther for parts and supplies we used to get here in Mott. But there's some stability today with CRP. It helps many producers by giving them a consistent source of income to supplement the cash flow of their farming operation. CRP has also taken a lot of highly erodible land out of production.

"Because CRP has helped create habitat for pheasants, we're seeing a change in land ownership in Hettinger County. We are getting an outside presence from investors looking for hunting land. The last land sold here went for $450 an acre. The fellow who bought it only wanted it for hunting. Every year there's a new story about someone buying land for a premium just for pheasant hunting. Why not sell? Land went into CRP for a reason. It wasn't productive and now a farmer can sell it for twice what he paid for it.

"No question about it, the economic impact of pheasant hunting has strengthened local businesses. Some of our businesses make ninety percent of their income for the year during the ninety days of the bird-hunting season. The economic impact for Mott has been estimated at one million dollars and that probably could be tripled when considering the entire county. Most of the land around here is posted and fee hunting is common. Landowners are charging $150 a day to hunt. Some guys are making $10,000 during hunting season for letting people on their land. Over near Regent farmers have formed a cooperative, the Cannonball Corporation, to set aside more than 20,000-acres for private hunting.

"In a way the Buffalo Commons idea is coming true here. There's been a lot of marginal land converted to CRP and this has provided us with good hunting and eco-tourism opportunities. Our community continues to age and I wonder when the outmigration of our young people will end. But whether or not the Buffalo Commons ideas stand the test of time really doesn't matter to me. We will always have a population here dedicated to raising food to feed the world." ▪

John Fiedler stands in CRP land that has become a bonanza for hunters in the Mott area. Once only marginal cropland, it is now prime habitat for pheasants. The ninety-day pheasant-hunting season pumps millions of dollars into the local economy.

Marilyn Hudson looks at a photograph of the ranch her family moved to after Garrison Dam was built. Hudson remembers raising big gardens in the fertile soil of the Missouri River bottoms. By contrast, their "on top" ranch was dry and poorly suited for gardening.

Marilyn Hudson

Farming Before Garrison Dam ▪ Parshall

"I can see parallels with what our people went through and what farmers are experiencing today. We are both so dependent on government-funded programs for our economic existence. And we both know we have to diversify in order to make it, or find something else.

"My father was Hidatsa and Mandan, and we lived with about 120 families at Elbowoods. We were agricultural people. We had big gardens and we raised almost everything we needed. The Mandan, Hidatsa and Arikara had lived for centuries along the Missouri River, so we knew how to farm. We raised a lot of things: beans, squash, many varieties of corn, probably nine varieties – some for flour, some for grinding or storage. We always kept large stores of beans and corn for fear of crop shortages. Especially the old women, I remember they had cream cans filled with beans and corn. We had nice homes, with windmills to bring up water. We didn't have electricity, so everything was done by hand and we worked hard. We worked all summer to prepare for winter. It was a good life living on those bottoms. The soil was good and we had huge gardens.

"We were self-sustaining agriculturists. Nobody had much money, but people bartered with each other, so we didn't need much money. Everyone had a variety of farm animals. Chickens, pigs, cows. That's the way it was pretty much all over the state in the 1940s. Our standard of living was on par with everyone else. The government had parceled out the reservation into allotments. From the Dawes Act adults were given 160-acre or 320-acre tracts. We lived on Old Dog's allotment. We had 160-acres on the bottoms and 320-acres on top.

"I was seventeen when we left Elbowoods because the water behind the dam was rising. The dam had been a long time coming, so people's reaction to it was very gradual. When the government told us in the 1940s that it was coming, we opposed it. But we were told we either had to sell out or the land would be condemned and taken. We asked the government to put the dam farther upstream, but it needed to be where it is for enough flow to generate power. So in the early 1950s the appraisals began. There are lots of documents that talk about the value of our land, the timber, coal, hunting and so on. Each allotment was appraised. Old Dog's allotment was appraised at $14,190. It was hard for people to place a value on land that had come down from their ancestors.

"Everyone had a plan for moving on top. My father had Old Dog's pasture land on top where he kept his horses. It was north of the river, so that's where we went. The land was hilly. It was just grazing land. It wasn't good for crops and it didn't have adequate water. Other people went to Twin Buttes, White Shield, Mandaree or New Town. There weren't many schools around then; so many families went to places like Parshall because there was a school. People got scattered. Some went where they had relatives or owned some land or could get some land.

"Our lives changed, in fact our whole way of living changed. We had to move where no one had been living before, and it was dry up there. The Three Tribes had organized a revolving credit program and it was very effective in helping getting Indians started on a small scale. We had the Boss Farmer then. He was like an extension agent today, but he took an active role in helping teach us ranching or new methods in farming. The Boss Farmer also helped Indians arrange for credit to borrow money for seed, cattle and implements. For the most part they were really interested in helping us do well.

"The dam really ended our Indian style of farming we had known on the bottomlands. It destroyed the agricultural economy of the Three Tribes, there's really no doubt about it. We found it was hard to start over, especially on less productive land than what we had been used to. We also didn't have the financial resources to start over. We hadn't been living in a cash economy, it was mostly barter, and we needed credit to start over and it was hard to get. Most of the people couldn't make it on the land they had to move to. We quickly went from a sustainable lifestyle at Elbowoods to the deterioration of our society in probably ten years. Many families gave up on their land and moved into town so the kids at least had a school to go to. There were few jobs, so they had to take welfare. Most came into town with nothing to do.

"There are very few Indian ranchers on the reservation today. Maybe forty and they are minimally successful. They're barely scratching out a living from the land. It's very hard to make a living in this harsh climate and bad economy. One problem is that we don't own enough land. All those original allotments have been split up among heirs. My father's 160-acre allotment is now owned by nine of us kids, so there's not enough left to do much with."

"We have a recovery fund to offset damages from the loss of land to build the dam. We share in the benefits from power generated by the dam. Our tribal government wants to use this money to develop agriculture, but also to diversify into manufacturing, eco-tourism, buffalo and so on. We want to build a Mandan village for tourism, and also to reintroduce our gardens and the crops we used to plant. Indian people have a high rate of diabetes from eating all this processed food. So we're trying to go back to the foods we used to grow. It would be healthier for us." ▪

Marilyn Hudson is the administrator of the Three Affiliated Tribes Museum at New Town. She was raised at Elbowoods and moved with her family to Parshall at age seventeen in 1954 when water began filling the Missouri River bottomlands behind Garrison Dam. She is a graduate of the Haskell Institute in Lawrence, Kansas.

John and Mike McDougal discuss combining their wheat crop with Phil Soukup, who helps the brothers harvest their crops. Successive wet years have put a strain on their farming operation, and without a new farm program, the McDougals believe many producers around Rolla will continue to only break-even or lose equity in their farms.

Mike McDougal
Grain Farmer ■ Rolla

"Times here have been better. In 1992 we probably had the biggest crop this area has ever seen. We almost had two crops in one year. But on June 11, 1993 it started raining and it seems it's been raining ever since. Last year, 1999, topped it off. We didn't even get all of our crop in. When I say we've been flooded, I don't mean everything is under water; but our fields were pure mud and we couldn't get our equipment out there to plant. My brother John and I got just a little less than half our crop in. Southwest of here there were guys who couldn't plant any of their land at all because it was so wet.

"Old-timers can't remember another time like this, with so much moisture. There have always been cycles of wet and dry years, but the wet years haven't gone on this long before. There are some interesting stories going around about farmers whose fathers are still alive. Some of the older guys were getting tired of their sons complaining about not getting into their fields to get their work done. So a few of these old guys decided to get on the tractor and show their sons how to do it. I heard one of these guys was back in the house about fifteen minutes later after he'd gotten the tractor stuck out in the field.

"Because of all this excessive moisture, this area has been hard hit by scab. It's a fungus that attacks the head of the grain and drags down the quality of the wheat. Dockage – the wheat seeds, shriveled kernels, chaff and other materials that are separated from the grain when a farmer brings it to the elevator – is subtracted from what a farmer gets paid. Normal dockage for wheat around here is two or three percent. But with the scab damage we've suffered in the last few years, dockage has run fifteen to twenty percent.

"A farmer can weather the ups and downs with prices for a while, but not for this long. We have to rely on disaster payments and crop insurance payments. No matter what you hear about it, nobody can get ahead on that or even stay treading water for very long. Farmers in this area are losing equity every year. It's our biggest problem. There aren't many springs when farmers go to seed their crops and their bins are empty because there is no grain there from the previous year. When the banker tells you the best you can do is break even, or to cut your losses as much as you can, why do we go ahead and plant a crop? What would another businessperson do if break even were the best they could do?

"With so little grain being grown in this area, the local economy really takes a hit. It's especially bad for the grain elevators. They've survived so far on hope – hope that we will get a crop this year. If we could just get some clear, sunny weather and a warm wind to dry things out. Besides the moisture, another reason less grain is being grown around here is CRP. The CRP program has taken a lot of land out of production and it's helped devastate our small-town economies. There are schools, churches, even whole towns that are dying and it's not a pretty sight. Everytime we cut production in this country, somebody somewhere else in the world steps it up.

"I think the days of production agriculture – where we produce as much grain as we can and haul it to the local elevator to sell it and ship it to God knows where – is a thing of the past. We need to process it ourselves and market it ourselves. Why, for example, should John and I ship our feed to the West Coast when we could either be feeding it right here or turning it into milk through a dairy? It just makes sense to process it ourselves and keep the money here. Cooperatives like Dakota Growers are a step in the right direction. I hope the premium beef project gets going again. I really think these kinds of ventures are the future for North Dakota. The problem is, it makes things extremely tough when farmers don't have extra money to invest in these projects.

"We're trying to start a wheat-straw processing plant here in the Rolla area. We're back to an old basic rule of agriculture, 'wealth comes from the ground.' Without a doubt, straw is new wealth. The kind of company we are looking at would generate an annual payroll of over $2 million and the pay out to farmers for their straw would be around $4 million. That would be a $6 million return every year to this area. Money like that usually gets spent again and again as it circulates through the community. But we farmers don't have the money to fund a plant by ourselves, so we're looking for outside money – somebody that wants to invest in and build a processing plant. It's tough to attract investors when they might earn a twenty or thirty percent return in a mutual fund. Why would they want to risk their money on something like this? It's been a long road and we've suffered setbacks. But I'm an optimist and I believe it can be done." ■

Mike McDougal is a partner in a cattle and grain operation with his brother, John. Mike graduated from Jamestown College and taught at Glenburn before returning to the family farm. He and his wife, Bonnie, have three grown children. Mike is an advocate for value-added agriculture and is on the board of a wheat-straw cooperative being considered for the Rolla area.

Ed and Bunky Nistler

Back Home Again ▪ Beach

After Ed and Bunky Nistler sold their farm equipment and livestock and got out of farming, they still hoped to live and work on a farm. Ed hired on at a large irrigated farm in southeastern North Dakota. A registered nurse, Bunky worked at a clinic, nursing home and greenhouse. But after three years of homesickness and unhappiness, the Nistlers found new jobs. Ed now works at an implement dealer and Bunky is returning to health care.

"For us, working for a large farm was an opportunity to stay in farming and continue to have that lifestyle for our family. We're better off financially now than we were before we left Golden Valley County, but for us, mental well being was more important than financial well being. We gave it a good shot for three years, but the pull to come back home was too great.

"Most of the hired people we worked with were just like us. They wanted to stay in farming, but didn't want the headaches of running their own place. We worked at a true corporate style of farm. It was big, about 15,000-acres with irrigation and crop diversity. The two brothers who owned the farm were the CEOs and us hired guys did all the actual labor. Without a doubt, that's the direction farming is going, but we found out it wasn't for us. We didn't like it. When you have the love of the land in your blood you can't get rid of it, and that's us. In the future there will be a lot of misplaced small farmers working for big farmers somewhere else. You can see it all over the state. We were part of it.

"Working for someone else took a lot of pressure off. We used to worry about bills, the weather, and all the uncertainties of farming. For once we didn't have to worry about going to the bank for another operating loan and wondering how we'd pay it back. That part of farming was a load off our shoulders. We looked forward to getting paid an hour's wages for an hour worked. When you're on your own, you can work every hour of the day and not make any money. But we found out the price of happiness is important, too. The mental stress of missing what we had left behind was the hardest on us.

"We were used to working together as a family and that's what we missed most. We missed the wide-open spaces. We missed people just stopping over for coffee. We missed having animals. We missed the way of life people live out here, the unconditional friendships we had. We were afraid we had lost a way of life we might not ever have again.

We farmed for twenty years. We struggled, but we feel lucky our kids grew up on a farm, working alongside us. It's a way of life you can't put a dollar figure on. We had the chance to experience nature close up. We grew to be closer as a family and we shared everything. That's something we will always be grateful for. We worked together and did things together. Our kids learned the work ethic and the responsibility of taking care of animals. And you can't put a price on wonderful friends and wonderful neighbors.

"The 1980s just killed us because of drought, grasshoppers, debt and high interest. The job on the big farm offered a good hourly wage, health insurance and a 401K plan. It sounded like a good move because farming wasn't paying off for us. We were both working jobs off the farm to make ends meet. We'd leave notes for each other. It just got to be too much. We decided any money we made off the farm was going toward family living and if the farm didn't make it, so be it.

"Most people not connected with rural life don't understand how close to economic disaster most farmers are today. For us, it seemed like we were always three days from disaster. One bad year would wipe out five good ones. Nobody knows how close to ruin farmers live until you've been there and done it. It used to be, if you were willing to work hard, you could make farming work. But that's not true today. It just isn't there. The prices are so low there's no way you can make ends meet with rent payments, machinery and inputs. There's just no way.

"When we worked for the big guys we didn't talk much about what we had been through because we didn't think people would understand. For some people the economy has been pretty good. So many people have no idea what it's like to struggle. They have no idea what it's like to work three or four jobs just to make ends meet. We started from scratch and worked hard for everything we had. Working for the big farmer meant we could pay off some debt, but we almost lost our sanity because the change was so great. Even though we were getting ahead, we weren't happy. As time went on we knew family and friendships were more important that just earning good money.

"Farming now is a vicious circle. You don't know how to keep going or where to get off. The margins are so slim you're facing disaster all the time. We made our decision to get out and literally suffered through an auction and the sale of our cattle. It really hurt. Then, after we had sold our piddly little farm, Uncle Sam told us we owed income taxes from our sale. So now we're paying off that debt, too. See? You can't afford to stay in and you can't afford to get out.

"There no way we're going back into farming. We might run some livestock on the small amount of land we still own, but we're not farming again. We don't want to be working for the bank, the implement dealer, the fuel dealer and everybody but ourselves. You just keep going around and around on a merry-go-round that never stops. We have no desire to get back into farming. No way." ▪

Bunky and Ed Nistler are happy to be home in Beach after a three-year experiment with being a hired farmhand on a large operation in southeastern North Dakota. They each worked long hours and paid off many bills, but the lure of family and friends brought them back to the familiarity of western North Dakota.

Mike Warner checks a sugarbeet field near the Hillsboro beet processing plant. Warner was one on the initial investors in American Crystal Sugar and was the cooperative's legislative chair for many years. He is now very active in a spring wheat processing cooperative.

Mike Warner
Sugar Beet Farmer ▪ Hillsboro

"Compared to food and energy, in terms of diplomatic as well as economic importance, everything else in our economy is an afterthought. These two issues are what cause governments to rise and fall, and can destabilize peace of the world. This is the big leagues, and it makes Microsoft look like the corner drug store. What I'm talking about is the battle for American agriculture. We have to realize the rest of the world is not willing to leave food and energy to the whims of the free market. They simply won't allow it. As much as we believe in free trade and the notion that capitalism spawns an ever-growing economic pie that improves the lives of all participants; the vast majority of the rest of the world does not believe it or there is a despot who won't allow it."

"Our system is the best, but until we can get other countries to come over to our way of thinking, we'll have to use confrontation. We'll have to fight the good fight until we win. Why are farm prices so depressed? Our number one competitors, the Europeans, are not only producing for themselves inefficiently, they are taking the excess from their inefficiency and dumping it on the world market. This depresses world prices for commodities. European consumers are paying a tremendous price to protect the inefficiency of their agricultural system.

"This is what the Europeans are doing in sugar, wheat and durum, all very important commodities to our region. Sugar is often cited as the anti-free trade program in the U.S. when actually we are fighting a trade war with Europe for freer trade, and have been for more than thirty years. It's the one commodity that U.S. farmers don't produce as much as our country consumes. So we import some sugar. Because sugar is a deficit crop, we use import limitations as a way of supporting prices by limiting supply. If we open our ports to European low priced sugar, prices will collapse. We will, in effect, surrender our industry to the highly subsidized European industry.

"I believe the only way we'll win this economic battle is the way we beat the Russians. We tell the Europeans, 'You can't sell wheat cheaper than we sell it. We'll give it away before we lose a market to you.' We keep hammering away at them and drive them to the negotiating table with the overwhelming costs they're going to shoulder if they keep on underselling, dumping and depressing world markets.

"This requires flat out trade war policies and the dollars to fund them from Congress and the administration. This will require the sympathy of Congress to our cause. We don't receive the proper attention from the policy-makers now because we're not glitzy. We're this kind of mundane business out here that just keeps cranking along, and because of this we're treated as an afterthought. Our people in Congress are good people with the greatest intentions, but they need to understand that this is a serious international game being played by countries with their best diplomats and toughest negotiators.

"Ultimately, to have the support of Congress we need the support of our citizens. Do average Americas care about farming and world trade? I think average Americans, for the most part, are worried about their house payments, their children and their jobs. The rest of it is just passing interest. Even so, polls do show that farmers are lucky to enjoy a positive image in the eyes of U.S. taxpayers.

"But because farmers are so busy just trying to make ends meet and keep our farms from going under, we tend to under-fund and under-staff the organizations that represent our interests. This is why Congress gets a disproportionate amount of policy counsel from the large multinational trading companies, who invest heavily in influencing policy. The Cargills and Archer Daniels Midlands don't get together in smoke-filled rooms in Chicago to plan the demise of American farmers. These companies make their money by getting a percentage of profit on grain coming or going. They are responsible to their shareholders and are not especially interested in how many farmers manage to remain on the land. Sometimes their interests are convergent with ours. There are other times that require trade restrictions or punitive action toward our trading adversaries. This tends to restrict trade and then our interests can very easily be in direct opposition to the multinational companies. We often have the same ultimate objectives, but we disagree about how to get there.

"Another way for farmers to compete in world markets is to diversify. In my case I have tried to make my farm into a food company. That has meant investments in starting successful value-added cooperatives. These are businesses in which I can take the crops I grow and through the cooperative, add value to them. In other words, get more money out of them by participating in not just the growing but the processing and marketing as well. That's an exciting idea. It was that promise that enticed me to put aside my pharmacy career, return to the farm, and buy stock in American Crystal Sugar in 1972. Those of us who organized Crystal Sugar were people mostly of modest incomes and means who joined together for our common good. Our hopes and plans were just the kind of American success story that a naíve, young idealistic college boy couldn't pass up. Now we're doing the same thing with pasta and plan to duplicate our value-added success with the spring wheat farmers." ▪

After graduating from North Dakota State University, Mike Warner was a pharmacist until he returned to his father's farm to become one of the original investors in American Crystal Sugar Cooperative, where he served for twenty-three years in various positions including the board of directors. He worked on three farm bills and GATT trade negotiations during the 1980s. Recently he helped organize the Dakota Growers Pasta Cooperative and the Spring Wheat Bakers Cooperative.

Dr. Gerald Sailer
Rural Health Care Advocate ▪ Hettinger

After thirty years of being a physician and a leader in rural health delivery, Gerry Sailer retired to join his son, Ted, in a family ranching operation. Although he claims his primary interest is in breeding prime Angus cattle, Dr. Sailer is organizing the Good Neighbor Project to ensure access to healthcare for farmers and ranchers struggling through tough economic times. He is a Hazen native and a graduate of the University of North Dakota and Baylor medical schools.

"One of our challenges out here is distance. When I came to Hettinger in 1965 it was one of the most medically remote and underserved places in the United States. There was only a primitive triage system in place. A medical school classmate and I went to work, and two years later we managed to recruit another classmate here. Now we have fourteen physicians practicing in an urban-type hospital and clinic utilizing the latest technology.

"I've always felt rural people deserve high quality health care. We hit on the idea that satellite clinics would be beneficial to the people in this area because we realized just Hettinger alone couldn't support the technology we needed for practicing quality medicine. We were some of the first to try a satellite system concept and now it's fairly common. Today we operate eight clinics in the region.

"Because I work with my son now on the ranch, I spend more time with ranch people, and I realize how many are locked out of the health system due to economics. They cut back on expenses in order to keep the farm or ranch from going under. I've heard from so many people that there's no way in the world average farmers and ranchers can afford to pay health insurance premiums. So they drop their insurance, they stay home from the clinic and they do very little or no preventive care. It's a tough situation.

"What can we do? I've never been very excited about looking to the federal government for solutions. It's too complicated and restrictive. People around here are eager to help their neighbors; we know how to help each other in times of need. So that's how the Good Neighbor Project started. I thought we could find people who would put up money to help pay for the medical care of our struggling farmers and ranchers. One thing we need to remember, if a rancher goes under, that family has to move away because there aren't any other jobs for them around here. To keep our population from dwindling away, we need to do this.

"We've set a goal of raising $2 million and we have volunteers working on it. Everyone who is helping is enthusiastic about the idea. We're appealing to individuals, charitable groups and corporations for donations. Our West River Regional Hospital donated $200,000 to help administer it so all the funds will go to people.

"Our local people have stepped forward, but we've had a difficult time selling this idea to national corporations. Most of them say there aren't enough people out here who use their products. Their giving is tied to marketing. One veterinary supplier donated doses of vaccine for ranchers instead of giving money directly. It's disappointing more corporations haven't stepped forward. We've been able to help twelve families so far, which is barely skimming the surface. We make eighty percent of the health insurance payment up to $5,000 per person per year. Then the family has to buy deductible catastrophic insurance for $5,000 and above so one family won't completely wipe out the fund.

"In order for this to succeed, most of the funds will have to come from areas of the United States that are enjoying more prosperity than we are. One of the strongest attributes of rural people is helping each other, but the money just isn't here to address this need. I mean, our clinics cover four counties in South Dakota and five counties in North Dakota. We serve 20,000 square miles – more distance than five eastern states – yet there are only 20,000 people who live out here.

"When my son said he wanted to make ranching his life's work, I wanted to join him. When our family was growing up I didn't have much time to spend with them, and now we can make this a family operation. I got intrigued with this life from the patients I was seeing. You know, they never take a vacation, they probably don't have much of an estate, but they absolutely love what they're doing.

"It's been a steep learning curve, but I can say now we really enjoy the ranching life. We're running 600 cows and a yearling operation on 9,000-acres of our own land and some rented Forest Service land. When we started we agreed our first priorities were to be the welfare of the land, the happiness of our family and the welfare of our animals. After that we'd let the numbers dictate what we'd do. We put everything on computer and made some changes. We sold most of our hay and farm equipment. We graze all winter and need only about twenty percent of the hay we used to put up. We calve in May, which is much later than anyone else does, but it results in much more healthy calves. We use a rotational grazing system to preserve and improve our grass. We artificially inseminate our whole herd and sell heavier yearling cattle.

"It's just a different operation but it's based on arithmetic, and it's working. I didn't think it was possible and cattle prices haven't been that great until recently, but we've done well because we've reduced our input expenses.

"There are a lot of cultural things that make the rural way of life important to preserve. There's a certain morality and honesty that goes with living out here. Most business deals are done with a handshake. There's something about rural life that's good for people. This is a good place to live and it's a shame poor economics are driving people away." ▪

Dr. Gerald Sailer looks out over thousands of acres of grassland south of Hettinger where he and his son are raising Angus cattle. A pioneer and recognized leader in rural healthcare delivery, Dr. Sailer is leading an effort to provide struggling farmers and ranchers with health insurance - an expensive item when incomes are down from continued low commodity prices.

Eleazar and Herlinda Martinez
Migrant Workers ▪ Hamilton

The Martinez family has worked winters in Texas and Mexico and summers in the Grafton area since 1975. Eleazar and Herlinda Martinez were born and grew up near Monterey, Mexico, south of Laredo, Texas. They have been married since 1990. Eleazar works as a hired man for a Cavalier farmer, and Herlinda teaches at Cavalier Migrant Headstart.

"Today, machines do the work we've been doing for many years. The equipment on the farms keeps betting bigger. When we started out working for farmers in the Grafton area, most farmers had six-row planters. Now they are up to twenty-four row planters. On some farms mechanical weeders do what we used to do by hand. With machinery like this, it's no wonder they don't need as many people.

"For many migrant workers, the only thing they have known how to do is work in the fields. Either hoeing beets or picking potatoes. Most don't have much education. Hoeing beets is hard work. We often work twelve hours a day - from six in the morning until six in the evening, even when it's raining. Sometimes we have a good day and can quit at four. Some of the older people have been hoeing beets for nearly forty years. We admire them, but we don't know how they can do that for so long, especially when they get arthritis. What keeps them going? Their love for their families, probably.

"We used to have a lot of Hispanic people coming up to Grafton for the summer. But now it's getting so that many come up in the spring from Texas and Mexico for work but don't find any. So after maybe a week they get pretty edgy and frustrated. Then they drive off to Colorado or Idaho to see what they can get, or even go back to Texas. They just travel until they find something.

"Some migrants, especially the ones who are getting older, are pretty stubborn and don't want to go to school. So many of them, when they were children, moved around a lot and didn't get much education. Within one school year a family might work in Texas, Washington State, Idaho and North Dakota. Many of them would drop out early, sometimes even before the sixth grade. Now, they just want to keep doing what they've been doing. But then they find the people they've worked for before don't have work for them, so they have to go to a new place, to farmers they don't know, and ask for work. Or they need to find another kind of job.

"Migrants can always go back to Texas. Fieldwork is easier to get there because almost everything is done by hand. It's a lot cheaper for the farmers. But the pay is cheaper, too, and you can't put in as many hours. In Texas you usually work only eight hours a day, sometimes only three or four hours. You can't live on that. But in North Dakota, most of the farmers are good to their migrant workers. Besides, in North Dakota we feel free to leave our kids alone outside playing. In Texas we can't do that. People here are so helpful.

"We don't blame the farmers for the few farm jobs. If we were the owners, we would probably have to do the same thing. The big farmers are probably the only ones who can keep farming, but the low prices are hitting them, too. We've seen farmers we used to work for, who were wealthy and farmed a lot of land, who have had to give it all up and forget about farming. It must be hard for them, too, to leave the life they knew and go into something else.

"Eight years ago Eleazar's boss came up to him in the field and asked him if he would like to quit hoeing beets and do something else. Now he does all kinds of other work on the farm - planting, harvesting, spraying or mowing. He spends a lot of time in the truck or on the tractor. What Eleazar thinks of doing some day is driving a truck, one of those big eighteen-wheelers. He dreams of owning his own truck. Other migrants we know have gone to work for construction companies. Two of Eleazar's uncles own their own farms in Washington State.

"A lot of our young people have been getting more schooling so they can get better jobs. Many of Eleazar's cousins who came up here stayed in school and got their high school diplomas. They would work all day hoeing beets, then at night go to school so one day they wouldn't have to hoe beets. Now, one of them is an administrator in a hospital; another one is an FBI agent in Texas, and one is a physician's assistant.

"We have had a very good relationship with the farmers we've worked for. They are like members of our family. The farmers we work for now, the Mahar Brothers, are there when we need them. When we've arrived in North Dakota for a season without money, they've helped us out. When our kids get sick, they've driven us to the hospital. When Eleazar's grandfather died two years, they even bought him a roundtrip ticket to Mexico so he could go back for the funeral.

"Last summer Herlinda hoed beets for the last time. In June she completed her GED and will have her certificate from the Child Development Association very soon. That will qualify her to teach pre-school children. For the past two years she has been teaching at the Cavalier Migrant School. She loves little children. She put on a fiesta at the school, with a program of about twenty-five children singing and dancing, and dinner for their families. Last year she was elected president of Migrant Headstart for Minnesota, North Dakota and part of South Dakota. At the conference in Washington, D.C., this year, she was elected national treasurer for Migrant Headstart.

"We would like to see our two daughters get good jobs some day. Maybe in a school or hospital; maybe a business. But before that, they should hoe beets the way we have and their grandparents have, maybe for just a year or two. It would teach them to stay in school. It would teach them what life is all about and how to be careful with their money. They would learn the value of earning money." ▪

Herlinda Martinez talks with her husband, Eleazar, about her new responsibilities as regional president of Migrant Headstart. Both hoed sugarbeets in the northern Red River Valley for several years. Now Eleazar drives equipment for a Grafton farmer, and Herlinda, after completing her GED, has become active in educating children of migrant workers

Life on the ranch is fair game for a newspaper column by Dean Meyer. Any activity, such as arranging his tack room, may inspire an idea or a story. Meyer has worked on ag issues in the North Dakota Senate and led the unsuccessful attempt to establish a beef-processing cooperative.

Dean Meyer
Rancher, Columnist, Value-added Advocate ▪ Dickinson

"Did you hear the one about the farmer who asked the banker for an operating loan?

"Banker asked him what he intended to plant. Wheat, the farmer said. So he got the loan and planted wheat, but it wasn't a good year. He got hailed out. The farmer went back to the banker for an extension on another year's operating loan. Now he's two years behind. Again, his wheat crop was bad. Too much rain at harvest, it sprouted in the field and yielded poor. So he goes to the banker and asks for another extension. What do you plan to plant, the banker asked. Wheat, the farmer said. So he put in the crop but the price was bad. He went to the banker again. He said the markets were so poor he just wasn't making it and needed another extension. The banker told him to try planting watermelon, he'd heard the price might be good. So the farmer gave it a try and by golly, he had a great year. So he went to see the banker. Here's a check for that first operating loan, he said. Paid in full. And here's a check for the second and third year, too. Boy, the banker said, that watermelon crop was really going good. Yeah, the farmer said, and I saved enough money to buy my wheat seed for next year, too.

"That sorta sums up the state of mind some guys get into in this business. I think low prices are harder on grain farmers. It's probably easier for them to get out. I don't think you fall in love with a kernel of wheat. But a rancher is used to the day-to-day responsibility of caring for his animals. It doesn't matter what cattle prices are, you've still got to get up in the morning and go feed cows and do chores. You take care of that orphan calf and grain your horses. It gets to be your life. Nobody ever seems to get out of ranching; we pretty much die first.

"People need to be educated about our problems out here. When I was in the legislature it switched from a rural to an urban majority, and I think there is not as much understanding of rural problems now. It's gotten worse every session. I think state government can do something about our ag problems. There were a number of proposals last session that would have helped, but they were defeated. It's too easy to say, 'Well, farming is a federal issue.' It's just a good excuse to say it's too big a problem for one state to handle. But I think we'll see even more ag proposals next session because people are looking for solutions. The thing is, you have to try some new ideas even if they don't work out.

"I worked on Northern Plains Premium Beef for five years, and it was exciting and fun. We had all kinds of support and expertise. We had three thousand members at one time. But when it came time to sell shares, nobody had any money to buy shares. We didn't have anywhere to go from there. It was disappointing because people who are positive and know it's going to work surround you, but you're not getting input from the quiet guys who don't have the money to invest. We were learning on the go. We had meetings in five states and two Canadian provinces, and we thought nothing was going to stop us. But we only sold about a third of what we needed to start. Now there are a couple of groups doing exactly what we intended to do, and it's looking good for them.

"Our plan was to partner with the retailer on the far end of the production line. We wanted to guarantee him a product, and we wanted him to guarantee us a price. We would base it on the cost of production so we wouldn't get these wild swings in the market like we have now. We get two years out of ten where we make a bunch of money and then spend the next eight years losing equity. Sure, we would lose the high points in the market, but we'd lose the low points, too. The retailers were excited about it because they wouldn't have to go out and look for beef. They could set their menus a year in advance and know where their product was coming from and what price it would be.

"The reason our North Dakota ranchers couldn't buy any shares was because they had been losing equity for so many years they didn't have any money. We were talking a hundred bucks a share, but it's a tough deal when you're losing $150 a head because prices are so low. It's a tough deal to put together these new generation coops because by the time most people recognize they need them, they have very little money left and not much equity in their business. They've lost money several years in a row, and you're asking them to invest in a risky venture, and they just can't do it. In the years when prices are good, they don't need it and they're paying off debt they've accrued over the bad years.

"My column writing started when I was president of the North Dakota Stockmen's Association. After writing about everything I knew about cows in the first two months, I had two years to go with nothing to write about. So I started telling ranch stories. I wrote legislative columns for the local papers. Then the Water Users wanted a column, and pretty soon a newspaper called up. I couldn't believe they would pay me for them. Now I'm in half a dozen papers mostly in western counties. It's the highlight of the week for the people in the nursing home at Watford City. The columns are mostly real, maybe embellished a little. Very few are made up. Most are about what happens on the ranch, something that happens when we're doing chores or out riding. I get a kick out of it. I can't believe people like 'em; I get comments all the time from people, especially when I pick on Shirley.

"What is it about this ranching lifestyle that we like so much? Sitting in this house when it's ninety-five degrees and no air conditioning? But it cools down real nice in winter because we don't have a furnace. Every day there's a new problem, but the solution usually makes you feel pretty good. You know how it goes, it always rains right after a dry spell." ∎

Dean and Shirley Meyer relocated their ranching operation from Watford City after selling their ranch to the Three Affiliated Tribes. Both have served in the legislature; Dean for ten years in the Senate and Shirley is seeking her second term in the House. Dean was chair of Northern Plains Premium Beef that unsuccessfully sought to establish a rancher-owned marketing cooperative. He's now working to establish an equine center at Dickinson.

Chuck Suchy
Farmer, Singer, Songwriter ▪ Mandan

Chuck Suchy grew up on the small cattle farm where he lives today with his wife, Linda, and three children. He's has been farming full-time and writing songs about his experiences ever since graduating from North Dakota State University in 1971. He and Linda raise alfalfa and run seventy head of Angus cattle in addition to Chuck's performing about one hundred concerts a year.

"My posters say farmer, singer, songwriter and storyteller, it's all interrelated. I sell my CDs to get the money to raise cows to sell the calves to make the money for another CD. Music has always been a part of my being and my life. I've always sung, and being out on tractors for hours on end, that's how I spent my time – singing and daydreaming. Singing really didn't become an integral part of me until about the time I graduated from college and got married. Then it became a valuable source of income.

"Our farm really isn't big enough for making a living. Of course, I use the farm in my music. Maybe I've even over-romanticized some of the aspects of farming as a way to gain income from it. I don't think I've been dishonest. If it is over-romanticizing it's certainly heartfelt. And that's probably why I'm still here struggling with this place – because of my romantic notion of farming. If I were going to be only be a musician, I'd live much nearer a major airport to cut my travel expenses.

"*Dakota Breezes* is one of my best known songs. Its point of view is to experience the sensual aspects of going through a day of plowing. Starting with the dawn, it takes you through the heat of the afternoon and into the cool of the evening. It's about the relationship with the earth felt through the machine; and the appropriate relationship of the operator to not make the machine overpowering.

"I had the idea for that song for a long time, but to get it done I had to finally lock myself in the bathroom one day when I had the house to myself. I was in there for two and one-half hours. It's very energy intensive to put yourself mentally and emotionally in a place where a song can come out. I need to deeply connect with the idea to bring all those feelings to the surface. It's one song I feel I nailed, or at least all the ideas came through. I just got a royalty check for twenty dollars from Rounder Records for it, so I'm not getting rich on this singing deal.

"*Simply Fly* is one of my newest songs that's connecting with people. It was written in our front yard overlooking an alfalfa field. Watching these hawks ride the wind currents over the bluffs, I thought, man, I'd like to do that. I heard the words simplify, simplify, which turned into simply fly. I wrote a little poem about that experience and the music followed. The song is about longing for freedom and having less pressure in life, and the hawk says you already know the answer – simplify and simply fly.

"People tell me they like my songs because they express feelings for the farm and the rural life. So many farmers aren't able or are unwilling to express their experiences. Maybe it's part of the isolation of living out here, or the macho mentality. Some of it is denial. You know the strong, silent thing about not talking about what's happening to you. My dad was a fairly tough guy with great big strong hands. But what stands out is how soft and gentle and nurturing he could be, like helping a newborn calf start breathing. That's a very powerful emotion, so why don't we yield to it and appreciate it?

"There's a lot of pressure on farmers to succeed, and I think the image of the great American agricultural experience the media is feeding us is hard to live up to. If you try to live up to the ads in the farm magazines and the television commercials you put yourself at risk of failure or feeling like a failure by comparison. Farmers have to be strong enough to be comfortable with our place, our own operation, and withstand the pressure to be something else, because that something else is probably something we are not.

"As I grow older, my perspective on life has changed. I'm not so farm-bound anymore. For a long time I thought my life ended at that hill over there where our property ends. Anything beyond that simply wasn't for me. But my music has made me realize there's life beyond the farm. That's why I treasure my music so much, it's enabled me to see beyond the fence line. My songs have also grown and changed to where I'm writing about personal relationships and more universal themes than just farming. Unfortunately, that sophisticated world out there gets tired of only farming songs. Why people aren't more concerned about where their food comes from is beyond me.

"People come up to me after concerts and say my songs really hit home for them; my songs are about their lives, too. I feel good about that and honored, but at the same time burdened, somewhat, because I can only describe farming and its problems, I can't fix them. I don't make a lot of money doing this, so those comments are always the supreme payoff. I get a lot of letters, too. They say, boy, you've lived it, you're not kidding. That validates me as a farmer." ▪

Someday I'll die / with trees and grass
On some hill I'll lie / it will come to pass
It pleases me / to let this body be
Where a west Dakota breeze blows over me
— *"Dakota Breezes"* © *Chuck Suchy 1984*

Simplify simplify
You know no more less than I
Simplify simplify
Spread your wings,
Simply fly

— *"Simply Fly"* © *Chuck Suchy 1999*

Chuck Suchy's music is inspired by his daily work experiences on his farm south of Mandan. He spends about a third of the year performing at festivals and concerts where people frequently tell him his songs reflect their own lives on the farm.

Blooming fields of sunflowers cast a Griggs County landscape in brilliant colors on a perfect mid-summer day. The beauty of the land is one attraction of living in rural North Dakota.

A time of transition

Sheldon Green

This is a time of transition in North Dakota. As a new century begins, we are witnessing the deterioration of our rural areas and the only way of life many North Dakotans have ever known.

Fewer farmers are working the land and on average they are older than they used to be. Many young people are heading off to cities and greater opportunity. Farms are becoming fewer and larger. Small towns are becoming smaller. The trends in rural areas are in decline, accelerating each year toward fewer and fewer, less and less. These are tendencies not likely to be reversed.

The simple fact underlying today's trends is there's enough food being produced in the world to feed everyone. So why is rural North Dakota struggling in trying to keep up in today's economy with food production, our most basic industry? Two reasons: Raw material producers are being ignored by political and trade policies; and high technology food production no longer needs a large labor force or a large number of farms to produce the amount of food we need.

St. Paul Pioneer Press farm reporter Lee Egerstrom has identified a paradox that lies at the root of this transition. The paradox, he says, is that the future may indeed be good for those remaining in agriculture, but not for rural people and small towns. The dying off of the rural way of life has little to do with food production. The trend is toward fewer and fewer producers. Economists have estimated that in a few years the United States will only need 50,000 farms to produce seventy-five percent of total U.S. production. Today there are about 250,000 farms still operating. As one farmer on the preceeding pages said, "It doesn't take a rocket scientist to figure out where we're heading."

Researchers Frank and Deborah Popper believe the Great Plains has been hardest hit by the downward trends in agriculture and population loss. Of all the Plains states, North Dakota is experiencing the severest declines, as evidenced by the continued outmigration of people. With fewer than 650,000 residents, North Dakota is getting down to the minimum number of people, cities and institutions our land-based economy can support in this new economic era.

The problem with the Great Plains is that it has always been hinterland — America's Outback. Plains states have been exploited since settlement days because agriculture is our chief man-made economic force. Of all the Plains states, North Dakota in particular has not managed to shed its colonial status for these reasons: we produce raw materials for others to refine; the price for our products is determined by outside forces often to our disadvantage; oil and gas is piped to industrial centers outside the region; and high transportation costs further impact prices due to our remoteness and distance from markets. Our major transportation and communication routes fail to unite us with other Plains states in which we share similarities because they run east-west rather than north-south. These factors contribute to our reputation of being isolated individualists, which hinders our political and economic clout in arenas where strong, well-informed, well-financed groups influence events. The net effect for rural people is substandard earnings and increased costs. It is little wonder why people living here harbor deep-seated feelings of resentment, insecurity and frustration.

A national study by the Rural School and Community Trust released in August 2000, found North Dakota to be "one of the most rural states in the nation, with half its population in rural places, and more than sixty-five percent of its public schools in rural areas. The report said rural teachers earn less in North Dakota than anywhere else in the nation; that almost one-quarter of North Dakota's rural students live in poverty and well over one-third of the state's rural adults have less than a twelfth grade education. The report listed North Dakota as one of ten states in urgent need of policy changes in order to reverse these disturbing trends.

Throughout the Plains, from North Dakota to Texas, there are many common threads in the landscape.

Most of the region is dryland with some irrigation; rural areas are sparsely populated with few large cities; there are great distances and limited economic, social and public services. Plus, farmers on the Great Plains must have a great degree of flexibility and an acute awareness of timeliness in order to successfully put in and harvest a crop as the seasons quickly change in a climate that is legendary for its unforgiving harshness.

Our country has become an overwhelming urban and suburban society that has largely forgotten its rural origins. In the not-too-distant-past, many non-farm people had agrarian backgrounds and knew firsthand the hardships of farm life. Today, however, farmers make up less than three percent of the population and many urban and suburban dwellers are several generations removed from the farm. They see themselves not as descendants of farmers but as consumers who benefit from low farm prices.

The thinning ranks of farmers and the continued exodus from the farm to the city means the majority of people are unfamiliar with problems in rural America, the Great Plains and North Dakota. The decline of the farm population has weakened the political influence of farmers in state and national politics, while the growth of consumerism and environmentalism has strengthened the hand of those who support cheap food policies or use of public lands.

Within the remaining farm community itself, there is widespread disagreement. Some see themselves as "agribusinesspeople" and identify more with economic analysts than with fellow distressed farmers. Some see themselves as small farmers being sacrificed in order to keep food prices low or keep foreign policy agreements intact. Some feel they have no control over their own lives. Others feel there is a lack of direction and unity in farm activism. Most recognize the fact that every farming and ranching operation is a highly individualistic enterprise, thus making it difficult to enact specific programs or to make widespread generalities. And some economists believe there are still too many farmers left in agriculture and a good shaking out will be beneficial to everyone.

The depressing economics and good and bad cycles of farming are enough to make people crazy. If farmers raise a big crop they are punished by seeing prices fall; if nature gives farmers a bad harvest, they have to sit by and watch prices rise with nothing to sell. To make matters worse, the risks in agriculture are numerous and largely uncontrollable. Farmers are at the constant mercy of lenders, markets and the weather. Furthermore, farmers must pay for twenty-first century technology with 1950s prices for their crops.

These have been tough times on the farm while many Americans have watched their incomes and stock portfolios increase as other sectors of the economy continue to grow. The handful of politicians and farm activists who are still talking about this transition have yet to formulate a coherent strategy or generate a sufficient unity of purpose to do much about the decline of agriculture. Try as they might, neither the policymakers nor farmers have succeeded in getting back to times of prosperity on the farm.

How did we get into this situation?

The last boom in agriculture was during the 1970s, when exports to Europe and Russia, helped by a devalued dollar, shot up. Wheat was seven dollars a bushel. Farmers paid their bills, bought land and new equipment, and prosperity hit Main Street.

But the following decade was characterized by inflated land values, continually falling prices, shrinking markets and rising farm surpluses, leaving farmers with heavy debit loads. Farm after farm went broke. Then came horrendous drought in 1988-1989 when many North Dakota farmers lost two-thirds of their crop. Ground moisture became depleted and didn't recharge until successive wet years in the late 1990s. Some farmers pulled through with crop insurance, disaster relief payments and by selling stored grain, but since 1989 prices have generally stayed at or below the cost of production.

The difference between the farm crisis then and today is subtle. In the 1980s many operators were forced to leave their farms or ranches because of economics – high interest and high debt. Today, farmers are choosing to get out for themselves, often to preserve what equity they may have left in their operations.

Small towns in North Dakota were hit worse in

Girls with matching outfits and matching bikes nearly match a fading Buy Dakota Maid Flour sign painted on the co-op grocery store in Regent.

Sharing news, deciding political races and forecasting the weather all takes place in many small town cafes. If a town lacks a café, the second most influential gathering place is most likely a church basement.

the 1980s and 1990s than previous decades because of consistently poor grain prices, drought, lack of prospects for off-farm jobs and the effects of people leaving communities after idling their land in the Conservation Reserve Program.

Fewer farms and fewer people, depressed small towns with fewer and smaller businesses, schools and churches, means our rural areas are losing their stability. There's instability in a family that worries about where the next paycheck will come from, so there is instability in the town, instability in the school and church budgets – the whole culture. Members of many small town churches, for instance, once were primarily farm families. Today, these same congregations may have retired farmers as members, but very few active farmers on their roles. Small towns don't die, their Main Streets do, so what's left are largely retired people.

Sheridan County in central North Dakota, which is not alone in this trend, has no doctor, hospital, traffic light or bus service. During the decade of the 1990s, the population dropped by twenty-two percent. There's barely enough people to keep a grocery store going. At Regent in Hettinger County, a cooperative grocery store serves the area, but suppliers are increasingly reluctant to deliver goods to rural places where there are fewer customers each year. Out in the country, many people stretch the definition of neighbor when the nearest inhabited farm is more than thirty miles away. The towns that don't have much of a trade area and don't have any primary government services are the first to go. It is widely believed in North Dakota that those towns under 2,500 people will struggle to make ends meet.

A comprehensive economic study released by the state in August 2000, found that despite the 65,000 new jobs that were created between 1989 and 1999, North Dakota's economy failed to keep up with the rate of national prosperity, and employment growth has slowed because of the continued outmigration of people. The report urged the state to support "value added" industries, but said money to invest in new ventures was "practically nonexistent."

Gloomy statistics suggest a state in decline. There are eighteen North Dakota counties with no more than four people per square mile.

Since 1990, North Dakota's net out-migration has totaled more than 37,000 people, about equal to the population of Minot. In 1999, the state's population dropped by 4,142 people to 633,666 and the birth rate declined by five percent from the previous year. Most of these losses came from rural areas. During the 1990s, five counties – Burke, Sheridan, Divide, Cavalier and Logan – experienced a drop in population of more than twenty percent. Furthermore, the percentage of North Dakota's population over age sixty-five is much higher than the rest of the country. Twenty percent of our population – mostly in small towns – live at or below federal poverty guidelines. New construction averages less than fifty dollars per capita when the national average is more than $850. On average, North Dakotans earn eighty-two percent of the national average for wages.

During the decade of the 1990s, the state's population peaked at 642,858 in 1996, but it has fallen every year since to finish at 633,666 in 1999 – a drop of nearly one percent through the decade.

The number of farms in North Dakota dropped in 1999 to the lowest level since 1910 when such record keeping began, with 30,500 farms, 500 fewer than the previous year. The average age of farmers in North Dakota is now fifty-three years old, and the percentage of farmers under age thirty-five is dropping steadily.

In 1999 the unthinkable happened when sizeable portions of land – even fertile Red River Valley land – stood idle. Fourteen percent of cropland in the state went unplanted. Some of it was flooded, but most of it was land unwanted by those still left in farming. The amount of land being farmed dropped by 100,000-acres from the previous year, to 39.4 million acres, the lowest level since 1950. Many of the crops planted were disappointing, including durum, which had the smallest harvest since 1993, spring wheat and barley the lowest since 1988. The value of nearly every crop dropped in 1999 by an overall total of twenty-two percent. The value of North Dakota's crop production was estimated at $2.1 billion, down from $2.7 billion in 1998 and $3.3 billion in 1996.

Net farm income has also taken severe hits. From 1998 to 1999 net from income dropped forty-two percent, from $3.7 billion to $2.9 billion, the

New Salem proclaims its pride for the region's dairy farmers – and hopes to attract tourists – with this gigantic statute of 'Sue' the Holstein milk cow that overlooks the countryside along Interstate 94.

third lowest per-farm net income during the decade. In 1999 the average North Dakota farm earned about $15,000 compared to about $25,000 the previous year. Crop income was off by twenty-seven percent and livestock output by seven percent.

Wise, old farmers used to counsel young farmers to expect only two, maybe three – if they were real lucky – good years in farming. The time in between was for hunkering down and riding out poor prices. Farmers interviewed for this book remember the early 1970s as one good period in farming, but they are still looking for the second one now, thirty years later.

Whether we are aware of it or not, we are part of an expensive superstructure that rests upon the soil and depends entirely on the gifts of nature for our existence, like an inverted pyramid. We hardy give it a second thought, but the whole enormous superstructure of American society balances on a tiny point – the point where fewer farmers with larger and continually more expensive machines and chemicals produce as much food as they can out of fields that are less productive than they once were.

The common thread in North Dakota history is the dependability of the undependability of agriculture. This fact has inverted our thinking. We seem to prefer hard times because we have more experience dealing with them. As Sentinel Butte rancher Bill Lowman observes, "It's the good times that will get you, not the bad times." The implication is to not spend too much and hunker down, because the good times won't last for long.

Richard Critchfield, an insightful writer with roots in North Dakota, once observed that America's urban culture is at stake if nothing is done about the problems in our rural areas. He believed all culture has a rural origin. Farming creates societies that work because they are based on the durable principle of extracting a living from a renewable resource – the land. "No society can get too far away from its rural origins and farming, and stay healthy," he wrote. "For urban culture to stay healthy, farms have to survive and people have to farm."

Critchfield believed that rural areas must remain viable and populated with people of ambition, otherwise our cities will become populated by people who are less social, less trusting, more greedy, more self-indulgent and less concerned with moral principals. For decades, bright, aggressive young people from rural areas have replenished our cities.

The shrinking number of farms and the decline of rural communities is reducing the traditional self-reliance that has characterized rural life. The transfer from being rural to urban people is fundamentally threatening our character. David Danbom, a regional historian at North Dakota State University, believes that of all the Plains states, North Dakota has best been able to retain values associated with its rural past, which are now at risk. "North Dakotans value community, church and family," Danbom says. "We prize independence and hard work, and value neighborliness and mutual help as well. We live for tomorrow rather than today, and we tend to be remarkably friendly to practical innovation."

European immigrants came here to avoid absentee landlords, overcrowding, famine, social unrest and to overcome their lack of family inheritance by claiming cheap land for farming.

To the first settlers, the prairies looked inexhaustible – unlimited space and inexhaustible soil fertility. Prairie topsoil was measured at twenty-five inches thick in places on the Great Plains. Each ethnic group brought its own habits, manners and intensive, European-style agriculture with them, including the year-in and year-out preference for raising a single crop – usually wheat. These immigrants were generally unreceptive to difference, diversity and other ways of doing things. They brought their ages-old habits of supporting city people through their own productivity and the continuous export of their best goods – be they grain, milk, meat, coal or their children.

But the first visitors to the region saw things much differently.

In the 1790s explorers and fur traders like David Thompson and Alexander Henry noticed this region was different from the mountains, lakes and forests they normally operated in. From their point of view, the Plains had limited resources, and further exploration in the region was curtailed.

In 1804-1806 Lewis and Clark noted in their

journals the generally harsh climatic conditions of the Plains. They concluded that agriculture would be risky in the region due to the general lack of rainfall and moisture, the nearly constant and damaging wind, hot summers and prolonged, cold winters.

But it was Stephen H. Long's expedition into the region in 1820 that really put the resounding whammy on the Great Plains. In his report to Congress, Long wrote, "In regard to this extensive section of the country between the Missouri River and Rocky Mountains, we do not hesitate in giving the opinion that it is almost wholly unfit for cultivation, and, of course, uninhabitable by people depending upon agriculture for their subsistence."

Long's map that accompanied his report designated the Plains region as "The Great American Desert," a name that would stick in people's memories for generations. Long's report confirmed the observations of previous explorers from Coronado to Lewis and Clark, but it did even more. It prejudiced the American public with the use of the term "desert," and it became a reality in the minds of people who considered the region as a place only to pass through or avoid entirely.

In the 1880s, rancher Theodore Roosevelt was a keen observer of the Bad Lands country around Medora, Dakota Territory. He loved the clean environment, forthright people and the open spaces of the prairie, and he perceived change more clearly than most men of the time did. While he may have worried about overstocking the plains, he didn't think farming held much promise.

In an article for Century magazine, TR wrote: "The country throughout this great Upper Missouri basin has a wonderful sameness of character – it is a semi-arid region. A traveler seeing it for the first time is especially struck by its look of parched, barren desolation; he can with difficulty believe that it will support cattle at all. It is a region of light rainfall; the grass is short and comparatively scanty. All through the region there has been sparse settlement. Now and then miserable farmers straggle in to fight a losing and desperate battle with drought, cold and grasshoppers."

The Homestead Act of 1862, by allowing 160-acres to qualified citizens over twenty-one years old, was the great political solution for settlement of the Plains. But the number of acres were too small to sustain a family, so in 1873 the Timber Culture Act was added, granting an additional 160-acres provided trees were planted and nurtured on forty acres. This was considered landmark legislation at the time, since it was commonly believed that trees brought rain and rain followed the plow. But even legislation couldn't make trees grow on the treeless Plains, and the act was repealed in 1891.

The Desert Land Act of 1877 allowed settlers to purchase 640-acres for the low price of $1.25-cents an acre, but it required that the land be irrigated within three years. The legislation was considered absurd, and abuses abounded. Many settlers merely poured a barrel of water into a furrow and called it irrigation. The fee didn't make sense either, since Homestead land was free but "desert" land cost a buck and a quarter!

At the North Dakota Constitutional Convention in 1889, government surveyor and explorer John Wesley Powell urged the delegates to enact larger land units, to designate some land for pasture use only, and to organize irrigation districts. Especially in the central and western portions of the state, Powell advocated farm units be no less that 2,560 acres, a size he considered to be the minimum number of acres in which to support a family-sized farm or ranch.

But Powell's ideas were not accepted. Delegates who grew up in humid regions of the country before moving to Dakota didn't understand the unique characteristics of the Plains. Railroad boosters who dominated the proceedings were eager to sell their land in order to pay off their investment, and preferred smaller, 160-acre units so more settlers could come here and eventually become customers for shipping their grain. The new state officials, desperate to begin collecting taxes to pay for the extensive infrastructure they were creating, sided with the railroads.

Powell left the convention convinced the effect of the Homestead Act would be to attract people ignorant of the conditions on the Plains who were sure to fail. Many Dakota settlers plunged into immediate debt to obtain land, buy equipment and seed, and begin paying taxes. Fortunately, climatic conditions were favorable during the first decade of the twentieth century, and many homesteaders stuck. By 1924 the average sized farm had

Farming is in a time of transition and rural lifestyles are quickly changing, but people will not completely abandon our rural areas for life in the city. North Dakota will not become a vast, empty landscape because our land is too productive, our farmers are too skilled at growing quality crops, and a continually growing world population needs a reliable source of food.

increased to 480-acres, and by 1969 the average farm had grown to 1,160-acres.

Other states in the region entered the Union on different terms, however. Texas, being an independent republic, had its own public domain policy that discouraged speculation because land was generally sold in large units – about 2,500-acres – which approximated the real minimum requirements for a successful farm or ranch. The average Texas operation was about the same size as what Powell was recommending for all Great Plains states. Texas land policy was ultimately an advantage for settlers because it discouraged those who were not truly agriculturists or those who couldn't afford to purchase land. Montana at statehood decided to keep hundreds of thousands of acres in public land, which settlers could then lease. This resulted in much lower initial investment for Montana farmers and ranchers, and many were able to lease sufficient tracts of land that enabled them to support their families and earn a return.

Many historians and economists believe the federal government should have adopted the Texas model for settlement because it was based on realistic expectations. Railroad boosters held their own with arguments that nature had shown her kindness to the Great Plains by putting few trees on the land. The Plains were meant for the plow, they said, and the land was plentiful and productive. Only will power and determination was needed to become rich! Incidentally, Texas and Montana are two states that Frank and Deborah Popper have seldom visited to discuss their Buffalo Commons ideas. In fact, no farm group in Texas has ever invited the Poppers to speak. The Poppers have visited North Dakota more than any of the Great Plains states since their buffalo-based, eco-tourism idea first appeared in 1987.

Plains settlers eventually made major adaptations in proven farming practices. Powerful tractors were developed and deep plowing was discontinued; tillage was only to loosen the soil and assist in weed destruction. Summer fallow as done to accumulate and preserve two years of moisture in the soil and remove weeds. No-till practices evolved from maintaining a cover of stubble from the previous year's crop to help prevent wind erosion and preserve moisture. Seed companies and land grant, research universities in the region developed new drought-resistant varieties of seed. This new technology – big machines, herbicide-ready seeds, quick growing crops, no till or minimum tillage – are significant Plains adaptations to help farmers produce enough bushels to earn a living from farming.

One unique adaptation was a farmer-led adventure into politics. The Nonpartisan League was organized to help farmers put up a good fight over low prices and high shipping costs. The movement was most effective in North Dakota, but also spilled over into Minnesota, South Dakota and Montana. Much of the farmers' initial anger was directed at Duluth terminal elevators that were mixing poor grades of wheat with premium grades, thus lowering prices for producers. Leaguers believed that they were somehow second class citizens, and forces over which they exercised no control manipulated their lives.

In 1919 an NPL dominated North Dakota legislature created such revolutionary endeavors as a state-owned bank, state-owned mill and elevator, hail and crop insurance, and a worker's compensation system. William Langer, who rose to political prominence with the league, later declared a farm mortgage moratorium and effectively used the state mill and elevator to force Minneapolis millers to raise grain prices in 1933 when he was governor. Through politics, the NPL was able to help all farmers in the region by wrestling some favorable prices during the drought years from the market manipulators and big millers. The NPL was a short-lived phenomenon, but the institutions it created still exist today, and NPL influence stood as a voice for liberal change well into the 1950s. The legacy of the NPL is that it promoted progressive ideas for government and gave the rural majority a chance to take control of their own destinies.

Trained observers have long recognized a far-reaching truth about this region. It is that "the Great Plains environment constitutes a geographic unity whose influences have been so powerful as to put a characteristic mark upon everything that survives within its borders. The failure to recognize the

The 1919 farmer-dominated legislature, the heyday of the Nonpartisan League, started North Dakota farmers in the direction of processing their grain to receive higher value. Prior to that, local elevators were virtually forced to pay whatever price the Minneapolis millers or Duluth shippers were offering for grain.

Two wheat fields, separated by a narrow access road, stretch to the horizon in Sioux County near Selfridge. Immigrants eager to begin their lives in the New World opened virtually all the land they claimed in their haste to plant King Wheat. Today a few highways bisect vast stretches of wheat fields as people in North Dakota's western counties must travel greater distances to reach service areas.

fact that the Plains destroyed the old formula of living and demanded a new one led the settlers into disaster, the lawmakers into error, and leads all who will not see into confusion." Thus wrote the influential historian Walter Prescott Webb in his landmark 1931 book, *The Great Plains*. Webb's writing at the beginning of the Great Depression continues to accurately reflect the condition of this region now seventy years later.

One root of our problems then and today is our unforgiving environment. The environment of the Great Plains has three distinguishing characteristics: it is a comparatively level surface of great distance; it is a treeless, unforested land; and it is a region where rainfall is insufficient for the ordinary intensive agriculture common to lands of a humid climate. The single most distinguishing climatic characteristic of the Great Plains is a deficiency of water and rainfall. This deficiency accounts for many of the peculiar ways of life in the region.

Another climatic feature that has had important economic and historical consequences for our environment is wind. Nowhere in the world, perhaps, has the wind done more effective work than in the Great Plains. Lewis and Clark noted the velocity of the wind in their journals, which seemed to blow hot and dry in summer and was accompanied by sleet, ice and snow in winter.

Lewis and Clark saw the Great Plains as people of European descent who would soon follow them would never see it again. When European settlers came to the prairie, they brought their technology, their religions and their social attitudes, but also their view of the environment traced back to ancient times and the age-old dependencies on nature. In their use of technology, these agriculturists managed to help create a myth about the land that nature is completely malleable and that humans can change it and improve it in any way we like to achieve whatever economic or social goals we may have. Farmers who have struggled through the last two decades of drought and excessive moisture understand this myth. So do bison producers, who once they return their land to native grasses for grazing, notice wildflowers and wildlife – jackrabbits, deer, songbirds, hawks and eagles – return, too. There is a feeling that something right and good is being returned to nature.

Traveling up the Missouri River, Lewis and Clark saw daily that nature is continually changing. This is a message we tend to forget. On nice days we tend to forget the days when sudden, severe storms blow over grain bins and lodge thick, perfect stands of wheat. We forget the blizzards that leave us stranded in our homes without electricity. We forget how harsh and unforgiving this environment can be. Arriving at a relationship between our environment and us takes work. The old way of dealing with environmental problems was to assume that, undisturbed by human activities, living resources were constant, and their environments were constant. But in fact, nature constantly changes and we have to accept the naturalness of change, especially when we make our living from the land.

There is great variability in climate. Ten times a year the temperature falls or rises forty-five degrees Fahrenheit or more within twenty-four hours. The nation's historic high and low temperatures – minus sixty-six degrees and one hundred seventeen degrees – have been recorded on the Plains. The daily wind average is twelve to fourteen miles per hour and wintertime Chinooks can cause flash floods of rapidly melting snow. Indeed, storms over the prairie are like squalls on the ocean – sudden, intense, destructive, but local. On the Plains, there is no "normal" weather, only averages to compare.

Our region is also vast grassland. Grass prevails only where conditions are unfavorable to more luxuriant forms of plant life. Our hardy grasses are suited to withstand conditions of excess moisture or excess drought. A display at the Manitoba Museum of Man and Nature shows a cross-section of native prairie grass. The roots extend nearly three feet deep, and the topsoil is a collection of rich nutrients. Native grass is certainly a species that was designed to survive on the prairie.

Animals and the first people on the Plains learned to adapt to the deficiencies. Buffalo exhibit characteristics that suggest unique adaptation to this environment. Webb says buffalo came nearer to dominating life and shaping the institutions of the Native Americans than any other on the land. Expanding on that thought, it has been said that the history of the buffalo and the effects of people of European descent on this species epitomize the general history of the relationship between environment and western-style development. Marked by the arrival of Europeans who believed that agriculture and buffalo cannot mix, exploitation and eradication followed and, almost too late, an awakening of conservation and professional management barely saved this great beast.

The incredibly rapid demise of the buffalo demonstrates the power of technology when put to destructive purposes. The eradication of the buffalo by hunters and settlers was the conscious and deliberate result of people who were opening up the land. They saw prairie grass and buffalo primarily as obstructions to be removed once and for all. Only rarely did they see buffalo or native prairie as resources to use and maintain. Because buffalo didn't fit into the European mold of farming and ranching, they had to go. The plow took care of the thick prairie sod so King Wheat could be planted. Today only rare remnants remain of the oceans of prairie grass that had once covered more land area than any other kind of landscape in North America – more than the great eastern hardwood forests; more than the northern forests of spruce, birch and pine; and more than the western deserts and mountains.

It often seems that the common impression about this region is that, before the arrival of Europeans, Indians had essentially no effect on the land, the wildlife or the ecosystem, except that they harvested small amounts that did not affect the "natural" abundance of plants and animals. We know the Mandan and Hidatsa successfully farmed plots of corn and squash along the banks of the Missouri and Knife Rivers for 700 years. Their methods were neither intensive nor overly harmful to the land. There is ample evidence that Plains Indians greatly changed the character of the landscape with fire, and that they may have had major effects on the abundance of some wildlife species through their hunting. Sometimes they were merely practical. Instead of harvesting berries from bushes a distance from the village where attack from a warring tribe was possible, nomadic Plains Indians merely hacked the bush down and dragged it back to their village where berries could be picked in relative leisure and safety.

Those who understand the Plains adapted to them. It has never been the Plains that adapted to humans.

Montana historian K. Ross Toole believed "one of the notable things about the Plains is that they have rarely been what people thought they were - neither as bad nor as good, as rich or as poor, as beautiful or as ugly, as wet or as dry, as cold or as hot. The region has almost always been underestimated or overestimated." The quick history of settlement is that from 1900 to 1920 immigrant "honyockers" flooded onto the Plains, plowed up the land, met with devastation, and then flooded out. Some of them, however, are still here – those who adapted.

Adaptation took a big toll. People had to change and they had to learn to live in a sparsely settled area with a harsh, unforgiving climate. Open range ranchers reduced herds from 20,000 or 30,000 head to 2,000 or 3,000 head at most. Cowboys became fencers, sowers and reapers of hay, and builders of sheds and corrals. Homesteaders gave up on 160-acres and enlarged their farms anywhere from 3,000 to 10,000-acres, depending on the size of their families and the availability of credit. Plains towns learned to be content with being small. Most of our towns sprang up along a railroad line, born because there was water or fertile soil there. Some are simply the result of topography; they are natural trade centers.

Our towns are incredibly similar. Toole says they are characterized by the number of bars per capita, by the number of grain elevators and implement dealers, a weekly newspaper, if there is a motion picture theater, perhaps a fairgrounds or rodeo arena, or a community hall (but never a civic center). There is most likely no manufacturing plant of any kind. Most jobs are tied directly or indirectly to agriculture. There are generally several churches, which in many places are used in lieu of a community hall. Sometimes the school serves this purpose. The epicenter of the community is usually the bar, café or church basement. This is where news and information is shared, business is done, social interaction occurs, weather is forecasted and politics is decided.

Distances on the Plains impact everyday living.

In summer North Dakota's landscape – like this fertile scene near Elgin in Grant County – comes alive with color from the abundant variety of crops grown.

The North Dakota landscape looks so productive and so familiar it is hard to think of it as struggling or failing. But what has been called a "comforting illusion of familiarity" masks fewer inhabited farmsteads, declining school enrollments and rural towns that are getting smaller each year.

The McKenzie County school district at Watford City is as large as some eastern states. Each school bus travels an average of 110 miles a day. Put in a single line, the bus route is the equivalent of driving from Fargo to Seattle and back — every day! Families in our western counties must drive hundreds of miles to follow their children's school activities.

Considerable time, money and effort are expended in just getting from place to place. The nearest doctor might be two hours distant. Implement parts may be available only in regional service centers. The largest universities are in the Red River Valley, which serves to concentrate intellectual, athletic and cultural activities along a narrow corridor that also produces most of the state's political leadership. There are few influential newspapers in the region, and the great distances hinder the molding of public opinion. Cultural activities are sparse. Community theater productions are occasionally attempted, or a community chorus is formed for a concert at Christmas.

Buried in nearly every American's heart – certainly every North Dakotan's – is an emotional linkage to the land. The bad news is this emotional image causes us to keep repeating centuries old myths about rural people and farms – that life is somehow more noble, more moral, more uplifting on the family farm; and that personal initiative and hard work can overcome incredible obstacles. But hard work will not enable a farmer to buy a new $150,000 combine when his old one is worn out and worth less than $15,000. The reality today is that family farms are substantial business investments, and decisions about operating the farm have to be grounded in good business, not emotion.

We've been told the way of life on the family farm is dying, but this death has nothing to do with food production – production continues on. There have been some excellent harvests during these bad years. In 1999, for example, some farmers left farming because of continued poor crops and poor prices. Yet a few miles down the road, another farmer was having his best year ever. It's hard to generalize the situation in agriculture – farming is an individual art form. What we do know is the changing economies of scale, new technologies and the play of new producers in the world entering our traditional markets influence price. Farms may go under or get bigger, but food production continues on.

Another aspect of this crisis is the notion that agriculture as it has been done is agriculture as it must be done. An extension of this idea is that technology and machinery can make land produce whatever we want it to — as if through sheer force of will and the right chemical or machine, we can raise any kind of crop.

Of the periods of life on the prairie, some believe the hundred years we have tried farming it will be the very shortest. In his interview, Joe Satrom, formerly of the Nature Conservancy, expresses the view that today's intensive farmers probably won't last as long as the Mandan and Hidatsa Indians. In Satrom's view, farmers are asking the soil to do too much in order to continue a high degree of production. Some farmers are beginning to think land is most productive when left alone. Dennis Sexhus of the North American Bison Cooperative told us that nearly 100 years to the month after his great-grandfather had first broken sod to start the family farm, he had returned the land to prairie grass for grazing buffalo. "And both of us," Sexhus says, "did the right thing."

Some land is best suited to certain crops. Some land is best left for grazing. Everyone knows this, but when commodity prices stay so low for so long, farmers are forced to produce whatever they can on whatever land they have in order to generate income. This has lead to another paradox in agriculture. Our agricultural institutions continue to promote increased production in their research and development programs, which causes our government to respond by hindering the effects of these new methods through production controls on acreage and then providing price supports when farm income falls. Think of it. We possess the ability to produce huge stores of grain, which government farm programs must restrict to avoid being flooded with grain, then this same government pays farmers for the deficiency in the subsequent low price they can get for their crop. This effectively maintains cheap food prices for American consumers, and keeps federal tax dollars flowing into rural states to help farmers stay on the land. These policies move in conflicting directions and their impact is being severely felt in rural North Dakota.

The great attraction of the Plains was the way introduced plants and livestock would thrive on the prairie. The promise was that all the immigrants had to do to realize a life just like the one they were leaving was to clear away the native vegetation and bring on the plow. Today the prairies not only look different from what the settlers found, they are different. In the last one hundred and fifty years nearly half of the original topsoil has eroded. This is in spite of the lessons of the Depression, advances in farming techniques and the development and influence of soil-conservation districts. The fact is, soil erosion is happening faster than ever.

Once the plow had opened the prairie, weeds moved in and thrived. In their rush to subdue the land by uprooting prairie grass and replacing it with crops, immigrants found that once again nature would not cooperate. Weeds frequently did better than the crops they planted. Weeds, writes one plant expert, "are the inevitable result of any human attempt to restrict large areas of land to a single plant." Farmers were virtually powerless to prevent the entry of unwanted species. Weed spores were blown by wind and took hold in the soil with unimagined determination and vengeance. Much of a farmer's time and money is spent fighting weeds, and the amount of chemical used to kill weeds is both expensive and worrisome for the impact on the environment.

One of the problems in agriculture today is the high cost of inputs to control weeds and fertilizers to overcome the loss of the prairie's natural productivity. Damaged land and less productive soil are not the only signs of intensive farming methods, but are evidence that those methods are inherently destructive and self-limiting.

The fictional character Sherlock Holmes was fond of saying, "You see, but you don't observe." This too, is part of the problem in agriculture. Young city people have little understanding about life on the farm, and usually have a romanticized image of rural life. Those of us who live here also don't completely see the problems because it is so tempting to continue seeing what is familiar. We see straight rows of corn and sugar beets, thick stands of wheat and blooming fields of sunflowers. Our state looks like a Garden of Eden. Everything looks so productive and so familiar it's hard to think of all this as struggling or failing. It's too difficult to understand change when we're still caught in our comforting illusion of familiarity.

The big question today is "Who will farm this land?"

Widespread coffee shop wisdom says that any day now, big corporations will take over farming. This is a myth. Corporations are generally reducing overhead and selling off all their unprofitable investments. No, the greater danger, which seems to be true, is for the big corporations to control the prices on chemical and seed inputs, and control the markets.

The independence and freedom to do anything one wanted on their farm is virtually over. Farmers have become very selective about what they plant today. If there is no market for it, it doesn't go in the ground. Farmers now produce exactly what the market is asking for in order to expect any kind of return. Many farmers sign contracts that guarantee a price to produce and deliver a crop for a corporation or a cooperative. This trend appears to be where some farmers are headed in the new future. The day of the independent operator, running his farm operation as he sees fit, is over. Global markets and bankers simply won't underwrite this lifestyle any longer. Another emerging trend is farming for wages. Farmers who once owned their own small operation are being paid to work on the big farms that are presently succeeding because of their scale.

"There's no question in my mind that contracting for crops is where we're going," says Hurdsfield farmer Norman Weckerly. "I have to think agriculture is going to survive, and people will still make a living from farming. Some of us will just be farming for someone else. Now, some will say contracting to raise a crop makes us serfs of the land. In a way that's true because we're under the control of someone else. But you never contract to lose money. You sign a contract because it's profitable for you."

Some experts say this country needs two farm programs. One to help farmers weather low prices, the other to help retrain farmers after they leave the

North Dakota's unspoiled natural environment may yet play a prominent role in the future economy as a tourist destination. City dwellers will delight in watching buffalo settling in for the night highlighted by a lingering August sunset, at Theodore Roosevelt National Park.

Agricultural historian Hiram Drache has long believed the Homestead Act was one of the most harmful public policies ever put on the Plains states because it didn't provide a family with enough land to earn a living. Farmers often started in a financial hole and stayed there through years of poor yields or poor prices.

land. Job Service North Dakota runs the Survival 2000 program, offering training assistance for farmers, ranchers, farm workers and family members. The program provides funding for retraining, in-state travel for job interviews and relocating for new employment.

Other experts say the United States needs to aggressively market our farm products to the world. The leading exporter of wheat in the world today is the European Union, which exports six times more than Australia, Argentina and Canada together. While the United States has deregulated its markets, Europe has adapted guaranteed prices, export subsidies and effective campaigns to buy European before foreign. Others say if Americans want to continue having cheap food, which now in effect is subsidized by farmers willing to stay in farming at break-even returns or less, the government should compensate farmers so they can afford to continue farming. In discussing plans for a new farm program to go into effect in 2002, Agriculture Secretary Dan Glickman departed from the policies of Freedom to Farm – the idea of getting farmers off the federal dole – when he said future federal farm subsidies should be targeted toward guaranteeing all farmers a minimum income instead of supporting the prices of select crops, as the government has done since the 1930s. Only seven states, which Glickman called "Depression-era regions" that include North Dakota and Minnesota, get half of all the government farm payments each year while accounting for only thirty percent of the nation's agricultural output.

Radical solutions, except for the 1919 NPL initiatives, have traditionally gone nowhere in farming. The radicalism of American farmers extends only to their adaptability toward new techniques to improve yields and use newer, bigger machines. Politically and culturally, farmers are conservative with an abiding respect for the traditional way of doing things. In other words, new ideas and solutions will continue to be slow in coming.

Former Governor George Sinner argues that farmers must enter the food processing chain in order to remain successful. Dakota Growers Pasta and the North American Bison Cooperative are two very successful examples of this kind of diversification. They are prime examples of new business structures called "new generation cooperatives" that have become popular with farmers and rural development specialists in the last decade. They operate like an investor-owned firm, requiring members to invest by buying stock in the cooperative and delivering grain or livestock to be processed. The goal is to maximize economic returns for members by adding value beyond merely making a market for member's farm and ranch crops and animals. About half of the New Generation Cooperatives launched in the past ten years have been in North Dakota and Minnesota. Rural development specialist Bill Patrie says these co-ops are popular because they offer farmers and ranchers hope, the one thing they can't find anywhere else.

But the fundamental problem remains unsolved – American farmers are able to grow more food than other rich countries are willing to buy or poor countries can afford. There are no villains in this new reality. It is only change brought on by better technology that allows us to produce so much.

It has always been a tough life on the farm and many farmers instinctively know the meaning of the old blues lyric "*you gotta keep on keepin' on.*" It is how farmers have managed to get by for so long – *by keepin' on.* ■

Source Notes
"A time of transition"

[Page 113]

Lee Egerstrom discusses this paradox in his book *Make No Small Plans: A Cooperative Revival for Rural America* (Rochester MN: Lone Oak, 1994), p. 198.

The quotation "it doesn't take a rocket scientist to figure out where we're heading" comes from an interview with Fred Kirschenmann, a leader in sustainable agriculture who farms at Windsor and Medina.

Frank and Deborah Popper detail these downward trends in agriculture and population on the Great Plains in "The Great Plains: From Dust to Dust," Planning 53.2 (1987):12-18.

Carl F. Kraenzel described the Great Plains in terms of "hinterland" and "colonial status" in *The Great Plains in Transition* (Norman: University of Oklahoma Press, 1955), p. 212.

The report by the Rural School and Community Trust on the condition of the state's rural schools appeared on p.1, The Forum, Wednesday, August 30, 2000.

The report on population loss in the 1990s came from the State Data Center at NDSU and appeared in the Wednesday, August 30, 2000 edition of The Forum, p.1.

The heavily symbolic concept of a Buffalo Commons is hard for many Plains people to accept, because the return of land to its natural state seems to signify the defeat of long-held dreams and a drift back to the past. There are many reminders of change visible in the countryside as old institutions give way to new methods, such as rural schools from homestead days that have sat idle due to the population shift from rural to urban.

[Page 117]
Data on population loss in rural North Dakota is available at the North Dakota State Data Center at North Dakota State University in Fargo. One source that traces the percentage of Americans engaged in farming over the years is Hiram M. Drache, *History of U.S. Agriculture and Its Relevance to Today* (Danville IL: Interstate Publishers, 1996), p. 466.

The statement that the past few decades have been worse for farmers and rural communities than the 1930s is taken from Mark Hanson, "State Loses Farmers," *Bismarck Tribune*, 27 February 1999, 1B.

Statistics on decline in North Dakota population have been taken from "North Dakota Loses an Estimated 4,000 Last Year," [Fargo-Moorhead] *Forum* 19 February 2000: C1.

Statistics on the decline of population in North Dakota counties appeared in The Forum, Wednesday, August 30, 2000, p.1.

The decline in the number of North Dakota farms has been taken from "North Dakota Farm Numbers Drop," *Grand Forks Herald*, 20 February 2000: D1.

Value of farms crops is taken from an Associated Press story "No Improvement Seen for Farm Exports in 2000," [Fargo-Moorhead] *Forum* 2 December 1999: C3.

Net farm income for 1998-1999 was an Associated Press story "N.D. Records Third-Lowest Per-Farm Income Drop in 10 Years," [Fargo-Moorhead] Forum, 2 Sept., 2000, p. A-10.

[Page 119]
The statement that we "are part of an expensive superstructure that rests upon the soil and depends entirely on the gifts of nature for our existence" comes from Jane Smiley's essay "So Shall We Reap" in *American Nature Writing*, ed. by John A. Murray (San Francisco: Sierra Club Books, 1995), p. 23.

Richard Critchfield writes about the rural-urban relationship in his *Trees, Why Do You Wait?* (Washington DC: Island, 1991), especially on pages 5, 11-12, 223-224, 242-244.

The quotation by David Danbom is taken from "The Future of Agriculture in North Dakota," *North Dakota History: Journal of the Northern Plains* 56.1 (Winter 1989): 37.

Jane Smiley discusses the heritage ethnic groups brought with them to the Great Plains in "So Shall We Reap," *American Nature Writing*, p. 17.

[Page 120]
Information on David Thompson has been taken from Malcolm Lewis, "The Cognition and Communication of Former Ideas about the Great Plains" in *The Great Plains: Environment and Culture*, edited by Brian W. Blouet and Frederick C. Luebke (Lincoln: University of Nebraska Press, 1979), p. 33.

Stephen H. Long's expedition is described in Carl Kraenzel, *Great Plains in Transition* (Norman: University of Oklahoma Press, 1955), p. 61.

Theodore Roosevelt's quotation is from "The Cattle Country of the Far West" in *Ranch Life and the Hunting Trail*, a 1983 reprint of *Century* magazine from the University of Nebraska Press.

The Homestead Act of 1862 is described by Walter Prescott Webb, *The Great Plains* (New York: Grosset & Dunlap, 1931), pages 404-428. The Desert Land Act of 1877 is discussed by Webb on pages 412-415. See also "An Adapted Land Settlement Proposal" in Kraenzel's *The Great Plains in Transition*, p. 295.

[Page 122]
Information on the size of North Dakota farms comes from the United States Department of Commerce Census Reporting Center at North Dakota State University in Fargo and from Population Trends in North Dakota by County 1880-1990.

The Texas model is discussed by Kraenzel, pp. 83-84, and by Webb, p. 398.

Discussion of the Non-Partisan League is based on "The Nonpartisan League: The Courage to Stand Up for Farms" by Larry Remele, in *Plowing Up a Storm: The History of Midwestern Farm Activism*. Nebraska Education Television Network and Nebraska Committee for the Humanities, 1985.

[Page 125]
Webb discusses early Plains conditions and land features on pages 10-46 of *The Great Plains*. Lewis and Clark's journals on wind velocity are discussed on page 22.

European settlers and their perspective on the early Great Plains comes from Jane Smiley's essay, page 17.

Variability in climate is discussed by Daniel B. Botkin in Our Natural History: The Lessons of Lewis and Clark (New York: The Berkley Publishing Group, 1995), pp. 46-47.

[Page 126]
Webb's statement on the prominence of buffalo as an influence on Great Plains culture comes from his *Great Plains*, page 44.

Botkin discusses today's prairie grass as only a remnant of the seas of grass that once covered vast areas of North America on page 259 of his *Our Natural History: The Lessons of Lewis and Clark*.

Indians affecting the ecosystem is taken from Evan S. Connell, *Son of the Morning Star* (New York: Harper & Row), p. 138.

The history of "honyockers" and adaptation to the Great Plains comes from K. Ross Toole, *The Rape of the Great Plains* (Boston: Little, Brown, 1976), p 24.

[Page 129]
Emotional linkage to the land is discussed by Lee Egerstrom on page 23.

[Page 130]
The appeal of the Plains because of the way pioneers could bring various kinds of plants and livestock to the new land comes from Smiley, p18.

Jane Smiley speaks of the "comforting illusion of familiarity" in her essay "So Shall We Reap" on page 25.

There is a spiritual dimension to rural life, and people know they are not in control of nature. This cross set in a pasture in McIntosh County is one of many reminders along rural roads that a higher power influences all things.

"Like his namesake, George was a gambler. Nobody could farm that country without being a gambler. One good year, with enough moisture, plus high prices in the fall - that was all it took to make up for six or seven years of failure. There were smart gamblers and stupid gamblers, but every North Dakota farmer was a gambler, and even the smartest one reached a point, every season, where all he could do was stand and watch what happened to his crop like a man watching the spinning of a gambling wheel constructed in Hell. When several good years came along in a row, he cashed in on his lucky streak and put his winnings back into the game, like any other sporting adventurer, by investing in new buildings, new machinery, more stock, more land.

"But when the good years came even farther apart than the seven promised in the Bible, perhaps he failed utterly. Then he watched the last days of the earth, while plague after plague was unloosed upon him, with the hailstones as heavy as cannon balls, and the great star falling on the fountains of waters and scorching his unrepentant head, and the grasshoppers as big as horses, with breastplates of iron. Then he stood in the midst of the ruin, smelling the smoke from the bottomless pit, hearing the echoes of the last thunder and the final trumpet blasts, and he did not repent of the work of his hands. He was proud of having played out the game, even though his name be blotted out of the book of life. He was broken-hearted and wounded with the kind of permanent wounds that only the proud sustain, but still he was proud. If he had it to do all over again, he would choose to gamble again."

– Excerpt from Lois Phillips Hudson's critically acclaimed Depression-era novel, "Bones of Plenty," set in North Dakota

Is production agriculture doomed because of high costs and poor returns? Some think America should buy wheat from other countries and instead concentrate its investments on crops and industries with higher returns.

Unwanted bread and beyond

James Coomber

What difference would it make to the economy and food supply of the United States if North Dakota simply stopped producing wheat, beans, beets, sunflowers, or beef?

Even the United States as a whole, some suggest, should not be in the farming business. Steven Blank, an agricultural economist at the University of California-Davis, argues that the United States should abandon the business of production agriculture for a major economic reason: the American economy has evolved to such an advanced level that other sectors of the economy, such as high technology and the information revolution, are likely to be more lucrative investments than farming.

It is not necessary for the United States to remain in production agriculture, according to Blank, because it makes good economic sense for us to buy most of our food from other countries. Thanks to new technology and modern transportation, we are accustomed to going to our supermarkets and buying apples from New Zealand, tomatoes from the Netherlands, and beef from ranches in Australia and Argentina. This transition from growing our own to importing what we eat will gain momentum, Blank claims, without any significant change in our food supply or in the prices we pay at the supermarket checkout.

We have surpluses of many commodities, and U.S. farmers are competing in a world market against farmers whose governments provide generous price supports. Growing our own food comes at a great price. In 1999 North Dakota farmers alone received $1.39 billion in government subsidies and other cash payments. The equivalent figure for 1998 was $860 million. For years one of the biggest sources of income in North Dakota has been government payments, with much of that going to farmers. Unfortunately, these disaster bailout payments only allow farmers to stay on the treadmill of contin-

ued low market prices for what they produce. The payments merely help our farmers to continue subsidizing the cheap food readily available in our supermarkets.

Americans tend to think of the past decade or two as low points in the farm economy. But Blank points out that in the history of American agriculture, farming has generally not been profitable, certainly not when government help is subtracted from the revenue earned. While agribusinesses – large-scale farm operations – seem made to order for the present competitive farming climate, growing larger is not necessarily the solution to low returns. Eventually even most large-scale farmers may find it difficult to compete with cheaper foreign food. While the prediction that "the future is likely to be good for agriculture but not for rural people or rural communities" might be true in the near term, Blank suggests that eventually most American agriculture will become unnecessary as we increasingly import our food from around the world.

Blank's proposals might remind many Great Plains people of the controversial work of Frank and Deborah Popper, scholars in demography and geography, respectively, from the New York City area. Based on demographic factors they had observed in many Great Plains counties, they predicted in 1987 the economy and population of this region would seriously decline. Factors they examined included both long-term and short-term population loss, population density in relation to availability of public services, median age of a county's residents, rate of poverty, and construction activity. Economically viable farms and ranches should continue to operate, they recommended, but for economically vulnerable areas of the Great Plains the Poppers suggested that with government help the land be returned to its natural state, a vast area which they called a "Buffalo

Commons." In addition to the buffalo being a natural part of the environment, these shaggy creatures could offer an economic alternative to dying Plains towns.

When the Poppers spoke in various Great Plains communities in the late 1980s, they sometimes encountered strong opposition from a number of people who heard their message as an assault on their livelihoods and even on their ancestors who settled here. But when the two were invited to return to North Dakota in February of 1999, they spoke to a large audience in Fargo and received a much friendlier reception. Clearly they had been on target about the decline of numbers of farmers and of many rural communities. They had also been correct in predicting an increase in bison, though that increase has come not because of government intervention but rather from economic development and producers' strategies for diversifying their farming operations.

Why farming and rural communities on the Great Plains are on the decline is easy to understand for farmers whose crops bring 1950s' prices in a world of twenty-first-century costs. Paul Thomas, a Cannon Ball-area farmer, noted in 1999 that he would have been pleased with three dollars per bushel for wheat, but nearly a half century ago, in 1953, his father was getting three dollars and fifty-cents per bushel. Dairy prices have recently fallen to 1978 levels. Many other commodities are also bringing rock-bottom prices. Despite increased mechanization and dramatically improved crop yields, North Dakota State University agricultural economist Won Koo concludes, "Farmers' net return per acre has remained essentially flat, when adjusted for inflation, over the period of 1930 to 1995." Rural people have long complained about the differences between what they receive for what they produce as opposed to what their commodities sell for in the supermarket. While Europeans spend an average of eighteen to twenty percent of their paychecks on food, the average percentage of take-home pay for Americans that goes to food is ten to eleven percent, including restaurant meals as well as grocery store purchases. Indeed, cheap food has significantly benefited the rest of the economy, enabling Americans to spend more on other items like travel, larger homes, newer cars, and more fashionable clothing.

One effect of poor prices has been the dramatic drop in the number of farms. Sixty years ago there were approximately eighty thousand farms in North Dakota, but today there are less than a third that number. Since 1930, when the peak number of farms was reached, the number of farms has steadily declined while the size of farms has increased. Just from 1990 to 1999, the number of farms that gross at least $10,000 dropped from 28,600 to 22,100, a decrease of twenty-three percent in only nine years. In the years 1992-1997 North Dakota lost ten percent of its farmers; in one county that figure was as high as thirty-three percent. To look at the problem from the perspective of net income: in 1984 the average net income from beef farms in the West River country was $12,510; in 1998, that average came to $1,056. For the state of North Dakota as a whole, median net farm income decreased from $31,603 in 1996 to $14,290 in 1998.

No wonder in various surveys a large number of farmers report that they plan to quit farming within the next few years and that even more indicate that they would like to. Many face a difficult choice: stay in farming and expect mounting losses, or get out of farming and take with them a heavy debt. It is easy to understand why so many farmers are in difficult financial straits. But it is also easy to understand why even those who are managing to keep their heads above water are tempted to leave farming for more lucrative careers. Various people we interviewed remarked that while in the 1980s it was largely the less successful or less competent farmers who were leaving farming, in recent years the farmers who are leaving include some of the best. That's why many farmers not only refrain from encouraging their children to go into farming but also frankly advise them not to. In the words of one Dickinson-area farmer who is also a parent, "It isn't fit for them to be out there."

The entire student body of the Connell School, a one-room elementary school in Billings County, displays their academic achievement awards. Sparsity of population is closing and consolidating schools, including ones in remote sections of the Bad Lands.

Farm families love the rural lifestyle, especially for raising their families because children learn responsibility and the value of meaningful work. Many families say they can no longer recommend farming or ranching as careers for their children because of the economic uncertainties.

As many younger people head for the urban areas and the average age of the rural population increases, demographic patterns and even the character of the state are likely to change. According to North Dakota State University professor of sociology, anthropology, and agricultural economics Richard Rathge, in 1960 about one-third of all North Dakotans lived in the state's four largest cities: Fargo, Bismarck, Grand Forks, and Minot. That proportion has grown to nearly one-half today. One effect of that migration is political, and it goes beyond the problem of rural areas losing legislative seats to urban areas. In the past many urban legislators had grown up on the farm and were likely to be sympathetic to farm issues, but that is not likely to hold true for today's city-bred legislators who might be several generations away from the farm.

More serious for both rural and urban North Dakota is the loss of people who migrate out of the state. For the past several years only three states out of the fifty – North Dakota, Rhode Island and Connecticut – have been losing population. Recent Census Bureau data show that from 1990 to 1999 North Dakota's population declined by 5,134 residents, a decrease of 0.8 percent of the state's total population. According to Curtis Stofferahn, a rural sociologist at the University of North Dakota, "We have an increasing number of counties that are frontier counties . . . The model we're looking at for North Dakota is the Australian Outback. We're going to have a large ranch with hired labor and no neighbors." The "outback problem" is especially serious in southwestern North Dakota, where it is projected that from 1980 to 2005 there will be a twenty-five percent decrease in population. Even Fargo, with its present growing population, is likely to face a decline in the future, for the city's growth has been generated in large part by an increase in middle-aged and older people who are beyond their child-bearing years.

There is always the lingering concern that the decreasing number of farms and farmers will concentrate land ownership in the hands of fewer, larger operators, eventually ending most individual ownership of productive farmland. Inability to own farmland has caused tumultuous political turmoil, including major revolutions in France, Russia, and Cuba, where a few had held the land at the expense of the majority. Land concentrated in the hands of a few also caused many of our ancestors to leave countries like Germany, Norway, the Ukraine, and Sweden to come to the northern Great Plains in search of cheap land and a new life. Precisely what attracted immigrants to North Dakota, former Governor George Sinner suggests, could be something their descendents could lose. Similarly, in 1999 Minnesota's Roman Catholic bishops asked that government policies encourage ownership of land by many rather than "return to the age of land barons and serfs."

The costs of a poor farm economy are not only economic but social and psychological as well. Morale suffers with the frustration many farmers feel when they have no control over their futures, when they seem to be victims of overwhelming forces. For producers to plant a crop knowing that they are likely to make little or no money from that crop despite all the hard work they put into it is demoralizing, yet that is what is happening all across the state. Even people who love to farm find it difficult to resign themselves to earnings more in keeping with fifty years ago, especially when they look from afar at a booming national economy and urban people earning significantly better salaries and enjoying a more prosperous lifestyle. Sociological factors apparently account for some outmigration from rural areas. Agricultural historian Hiram Drache points out that even in countries like Norway and Sweden that heavily subsidize farming, numbers of farms and farmers are significantly decreasing. From this he concludes that in both Europe and North America rural people are leaving the farms not only for economic reasons but also because, especially as farm population and community decline, many have come to want what the city offers.

Psychologist Val Farmer summarizes the feelings he has observed in many farm people: "I see a lot of people wearing out in farming. They are just staying even-staying on the treadmill. The common way of putting it is 'it's no fun any more.'" Farmer notes that those frustrations are often aggravated by poor communication. Not only are there fewer people to talk with in rural areas, but there is also a limit to how much the people who are there want to discuss their problems. At the local café farmers might freely join their neighbors over coffee in a general gripe session, but they are understandably reluctant to give details about their own difficulties. On the other hand, some prosperous farmers avoid talking about how well they are doing, not only to keep from depressing others around them but also to avoid being resented-an interesting variation on the price of success. One North Dakota farmer recalled how in the more prosperous 1970s he was so proud of his new tractor that he would park it out near the road and keep it clean and shiny. But today he said that he would more likely leave the new tractor splattered with mud and parked back behind the barn, out of sight. Carol Bly is one of many writers who have discussed how Midwestern reserve and reticence and the expectation that people keep unpleasant or controversial matters to themselves affects rural people. For both the successful and the struggling, the isolation that comes with denial, inborn reticence, preserving pride, or the refusal to discuss fears and failures for whatever reason likely results in feelings of powerlessness and loneliness and psychological separation from the community.

Throughout most of its history North Dakota's small population has been scattered over the prairie with a low average density of people per square mile. Since many rural residents have lived isolated lives, community and neighborliness have been important resources for life on the Plains. While we might expect that in times of crisis the rural people who are left would support one another, in her "Dakota: A Spiritual Geography," Kathleen Norris suggests that, surprisingly, one effect of the farm crisis can sometimes be distancing oneself from those who are having financial troubles. For some this is a way of protecting themselves, of denying that there is a crisis and that the

community is in danger. Several of our interviewees reported hearing remarks from retired farmers who wondered why, when they were able to "make it" in farming in the 1970s, younger farmers have such problems today. But another effect of today's outmigration from rural areas is a feeling of loss for those who have moved away. There is also a heightened sense of aloneness that comes with farms that are increasingly farther apart with fewer neighbors. Maintaining community becomes a challenge. Some complain that with fewer neighbors it becomes increasingly difficult to find enough people to serve on the church council and school board and to volunteer for community activities. Even in the more populated rural areas of the state, one can drive for miles on paved highways, passing numerous sagging barns and abandoned homes and windbreaks with nothing to shelter, and encounter only one or two other vehicles in an hour – mute testimony to outmigration and the erosion of community.

Yet, as bleak as the situation may sometimes seem, there is some good news and some reason for cautious optimism.

"There is life after farming," psychologist Farmer emphasizes. Many families who have given up farming have discovered a sense of relief with their decision and have adjusted very well to life off the farm, a new life they often find rewarding and less stressful. Various agencies, including Job Service North Dakota, have enacted special programs to help farm people make the change to non-farm jobs, and with favorable results. The strengths, mechanical abilities, skills, and attitudes typical of farmers serve them well in other careers, so it should not surprise us that many farmers have become successful bankers, social workers, counselors, teachers, clergy, insurance agents and salespersons.

Getting out of farming is not the only answer to today's farming situation. But while farming will probably be happening in North Dakota for many years to come, the farming of the future is likely to be a very different kind of farming than most of us have been accustomed to.

While size of farming operation does not guarantee success, especially when wheat is selling for

To reduce stress on the farm, mental health experts advocate rural residents should try to maintain a sense of humor about events beyond their control.

The Red River Valley sugar beet industry is an example of a highly successful value-added industry that was organized by farmers. The economic impact of jobs created by the industry is significant and vital to the North Dakota economy.

less than three dollars a bushel, the trend is clearly toward fewer and larger farms and fewer farmers. Some predict that the owner of a large farm will resemble the CEO of a corporation, spending more time at the computer than on a tractor. But even with this continued expansion of farms, some small specialty producers will likely continue to raise niche crops like carrots and organic produce. As for the family farm, few of our interviewees have even mentioned it.

Filling the needs of global markets will be one focus in farming in the years to come. Staying with the familiar can make sense and is certainly easier for many farmers, who may have produced wheat for decades just as their fathers and grandfathers had done. But farmers of the future, rather than expecting a market for every crop, must be prepared to produce a crop for a specific market. If Asian consumers want a certain kind of grain for their noodles, for example, world competition means that farmers will have to raise the durum that will satisfy their Asian customers. Taking a business-like approach to producing for the marketplace will make the difference between success and failure for many producers.

Farmers also will benefit from being not only producers but owners, processors, and marketers as well. Instead of delivering grain to the local elevator, for example, and letting large corporations cash in on much of the profit, farmers can join forces and process and market their crops themselves. Value-added agriculture has been successful for such cooperatives as the Dakota Growers Pasta Cooperative, the North American Bison Cooperative and for farmers delivering potatoes to a new processing plant at Jamestown. Not only do producers receive premium prices for their crop; they benefit from selling the finished product on national and world markets.

Keeping as much money as possible in North Dakota by producing bison, durum, potatoes, and other farm products makes good business sense for the state's economy and increases profit margins for farmers. The challenge for individual farmers entering value-added agriculture is coming up with enough capital from their operations to invest in a value-added venture. Access to capital has traditionally been a problem for North Dakota producers who wish to expand yet lack the necessary money to put their plans into action. Inability to attract capital has prevented a beef processing co-op and a flour-milling venture from starting up; several plans to build strawboard plants are on hold while investors are sought.

But value-added agriculture illustrates a trend that promises to be significant in the future. It is advantageous for farmers to work together. As St. Paul newspaper farm reporter Lee Egerstrom writes in his book on cooperatives, farmers in today's agricultural climate increasingly have two choices: to remain farmers and produce for someone else, or to become "farmers-businesspersons" and produce, process, and market for themselves as part of a cooperative. Since 1991 more than twenty-eight of these producer-owned co-ops have been tried in North Dakota, primarily because the cooperative model allows them to enter the value-added chain, which in turn offers producers hope for getting a decent price for their products. According to Bill Patrie of the North Dakota Association of Rural Cooperatives, who has helped start many new cooperatives, North Dakota is a great place to put "new generation" cooperatives to work because of certain characteristics our farmers generally share: farmers here are much more people-oriented than farmers in other states, they like people and can generally work well together, and they understand that to get things done, they must cooperate and share resources.

Another reason for optimism lies with world markets. Improved national economies could open new markets for countries that presently cannot afford commodities produced on the Great Plains. Economically healthy countries generally consume more food simply because as people's financial conditions improve, they can afford to buy more and better food. New markets continually appear for various commodities. Recently, for example, the Middle East has emerged as a potential market for sunflower oil. But the most promising possibility is Asian countries, where improving economies represent a potentially large market for producers of grain and meat. One source estimates that if the United States could gain forty percent of the Chinese wheat market, U.S. wheat exports would expand by nearly ten percent. In the spring of 2000 President Clinton, members of Congress, and trade negotiators were debating the possibility of offering China favored-nation trading privileges with the United States. At that time the possibility of freer trade with Cuba was also being debated in Washington. It is probably only a matter of time before shiploads of American products will be heading for Havana as well as for other parts of the world that have been closed to U.S. farm products.

Government programs offer at least some possibilities of hope. In the spring of 2000 the latest farm program, Freedom to Farm, was under attack from Democrats and Republicans alike, including some in Congress who earlier supported the program. Freedom to Farm was passed in 1996 on the assumption that farmers would be able to thrive in a free market. However, many countries protect their own farmers through heavy subsidies, which works against U.S. farmers in global markets. Since Freedom to Farm was initiated, Congress has passed three emergency aid payments to farmers to compensate them for loss of markets. In so doing Washington has not only helped many producers stay in business but also at the same time demonstrated the weaknesses in Freedom to Farm. Since farmers and farm organizations disagree broadly as to what a new farm bill should look like, improvements to federal programs to become more responsive to farmers will take time. One response has been Senator Kent Conrad's Farm Income and Equity Act, directed to the disadvantages U.S. farmers face from European Union subsidies. Conrad proposes that the government pay farmers export subsidies similar to what European farmers enjoy. Additionally, the Clinton administration hopes global trade discussions will convince European Union countries to decrease their agricultural subsidies.

Growing new or improved crop varieties also offers possibilities for farmers. Pinto beans have long been in many farmers' rotations, and recent reports suggest that our climate with its long, sunny days and cool weather is ideal for raising other kinds of beans as well. Genetically modified crops are another developing area of plant science that may facilitate raising crops on the northern that have never been raised here before. Seeds that have been genetically modified offer possibilities of superior crops with less dependence on pesticides and herbicides. A common example is Roundup Ready crops, whose genes have been changed so that the plant will not be affected by the widely used herbicide from Monsanto. However, development of genetically modified crops might be stalled by consumer backlash; in fact, many European markets do not accept any genetically modified agricultural products. Although some farmers are cutting back high-tech plantings, the future will likely see an acceptance of these innovative crops.

Even if plant science should enable us to grow crops in North Dakota that several years ago were not considered possible in our climate, could we raise those crops at prices comparable to those of other countries? Would it make more economic sense to import food from other nations who produce at lower cost? There are at least several problems with Blank's thesis.

If the United States were to largely abandon farming, we would be left to the vagaries of the international marketplace. One obvious problem would be the security of our food supply. Even with the food we produce in this country, we do not have complete security; problems such as bad weather or crop diseases can interfere in disastrous ways with price and availability. Adding to this vulnerability would be disturbances abroad, such as unstable political situations, less competent farming practices, and slow and perhaps unreliable transportation systems. Also, what if another nation fails to honor a trade agreement? North Dakota Farmers Union leader Robert Carlson recalls a soybean deal made with Japan in the 1970s. After the

Large fields of wheat straw left behind after the grain harvest, like this one near Hazen in Mercer County, have inspired many attempts by farmers to enter the strawboard manufacturing industry, thus adding value to a frequently discarded resource. The biggest problem, however, is North Dakota's traditional lack of venture capital.

agreement had been completed, the price of soybeans nearly doubled, and the United States opted to renege on the agreement so that our farmers could cash in on a better price. The world economy depends on agreements between nations just as between consumers and growers. Those who argue for more food imports cite agreements between nations as one of the best guarantees for peace and harmonious world trade relations. Yet the risk is always in times of shortage, when push comes to shove, that a nation might serve its own citizens first and out of expediency ignore agreements with other nations.

But another serious question that needs to be asked is whether nations of the world will indeed have the capability of supplying the United States and other countries with the food we need. World per capita grain production increased 2.6-fold from 1950 to 1984, outpacing population growth. But since 1984 those upward trends have reversed, with grain production failing to keep pace with population growth. According to a report by Lester Brown from the World Watch Institute, from 1984 through 1999 the per capita production of world grain declined by about ten percent. Decrease in world food production is affected by such factors as soil erosion and compaction; emptying of aquifers and soil salinity and waterlogging caused by long-term irrigation. Other factors point to possible grain shortages in the years ahead. First, there is very little promising cropland that is not already being farmed. Second, every day a considerable amount of farmland is turned to non-agricultural use. Third, since much crop production is already being carried out with modern technology, it is difficult to imagine how technology can be dramatically increased to meet population increases. Brown concludes, "Most future growth in grain output must therefore come from exploiting technologies not yet fully used." In May of 2000 the International Food Policy Research Institute concluded that nearly forty percent of the world's farmland is "seriously degraded," with the problem at its worst in developing countries-an indication that today's unwanted bread could be very much in demand tomorrow.

Wrapped into the arguments to secure our own food supply is our tradition of giving food aid to countries that desperately need it. We usually think of aid in terms of sending food, but many in the U.S. agricultural establishment also feel the obligation to share expertise with developing nations. Many North Dakota farmers, especially those involved in promotional groups like the Wheat Commission, have seen the food chain reach from a North Dakota farm to A starving family somewhere in the world. While Blank downplays our obligations to the less fortunate, in rural North Dakota there is a strong spiritual and moral sense of producing people's daily bread, even in difficult times.

Another reason for keeping farmers on the land and thereby keeping rural communities intact is a concept many Americans have ignored: dispersion of population. Farmers Union leader Carlson explains that European farmers are generally prosperous because their countries have supported the idea of having people not only living in large cities but also distributed over the countryside. Part of this thinking seems to be a strong desire to keep the countryside looking beautiful. According to Carlson, Europeans "would not put up with the sort of devil-take-the-hindmost buccaneer capitalist philosophy that we seem to have in North America, where we accept decaying small towns and abandoned farms." It might also be in our best interests to keep farmers and rural people living on the land rather than watching them leave for the cities. Even with the cost of government support for farm programs, is it possible that the costs are greater for local, state and federal governments to support people living in metropolitan areas than for an equivalent number of people to live in rural areas? Writer Richard Critchfield claimed that a strong agricultural base with a critical mass of people living on the land is necessary for any nation to properly function well, both economically and socially. Rural people, according to surveys cited by Vyzralek, tend to agree on the superiority of rural life as a character-building force as they suggested country people are "more industrious, bigger risk takers, and more honest, wholesome, religious, traditional, practical, and intelligent" than urban people. For whatever reasons, in the years ahead the American public and their representatives in government will need to decide how much

Global trade could immediately help North Dakota farmers sell grain to countries like Cuba and China, which stand ready to purchase as much as forty percent of their country's needs. Key members of Congress are also pushing to rethink or repeal the unpopular Freedom to Farm legislation, popularly called Freedom to Fail in rural areas.

Equality is still practiced on a small town playground where pick-up games include whoever wants to play.
Author Richard Critchfield believed "there is simply no substitute for the farm and small town when it comes to forming human culture." Critchfield wrote that the disappearance of farms and the shrinking of rural communities is reducing the traditional self-reliance that has characterized rural America.

they are willing to pay to support farmers and to maintain the infrastructure of farms and rural communities.

Farming will never be the same as it used to be, but there is little reason to imagine North Dakota as a vast, empty grazing land in our lifetimes. Such factors as a growing world population, strengthening economies in Asia and the Third World, and increased incomes of consumers in grain-importing countries might gradually increase prices for wheat and meat. A report produced by the Department of Agricultural Economics at NDSU predicts that Asia will import eighteen percent more wheat in the decade 1999 to 2009. Wheat exported by the United States could increase from twenty-five million metric tons to thirty-five million metric tons. These predictions are similar to expectations by the U.S. Department of Agriculture, which suggests significant increases for wheat, soybeans, corn and barley. Such reports offer hope for people who are willing to treat farming as a business and run their operations according to the demands of today's global markets.

No one who lives in North Dakota needs to be told that we are living in a time of change-for many people, difficult, heart-wrenching change. In light of what is happening in farming and rural life today, Fargo Forum newspaper columnist Jack Zaleski suggests that "the state has to redefine itself to conform to the obvious reality." That is just what the fifty people whose stories appear in this book have been doing. That is what we, too, need to do: "redefine" ourselves in light of the new realities of our land-based economy. ■

Source Notes
"Unwanted bread and beyond"

[Page 139]
Steven Blanks's arguments are found in his *The End of Agriculture in the American Portfolio* (Westport CN: Quorum, 1999).

Recent data on farm income in North Dakota and the amount of money farmers received in subsidies and other cash payments in 1998 and 1999 was taken from Jeff Zent, "Net farm income up in North Dakota," *Forum* [Fargo-Moorhead] 2 August 2000: B1.

Deborah and Frank Popper's paper that received such wide attention in 1987 is "The Great Plains: From Dust to Dust," *Planning* 53.2 (1987): 12-18. For more recent reading by

the Poppers about the Great Plains and their Buffalo Commons, see "Great Plains: Checkered Past, Hopeful Future," *Forum for Applied Research and Public Policy* 9:4 (1994): 89-100; "The Storytellers," *Planning* 62:10 (1996): 18-19; and "The Bison are Coming," *High Country News* 30:2 (1998): 15, 17. One less-than-favorable reaction to the Poppers is that of Frank E. Vyzralek, who argues that the demographers from Rutgers overlooked "the cyclical nature of the Plains environment" and assumed that the 1980s would be the norm for northern plains farming and climate conditions in his article "The Commons: The Life Cycles of the Great Plains Has Inspired Good and Bad Ideas over the Decades," *North Dakota Horizons* 21.1 (Winter 1991): 4-9. How public reaction to the Poppers has changed in the thirteen years since they had first presented their hypothesis is reflected in a Bismarck newspaper editorial, "A Decade Wins Them Credibility," *Bismarck Tribune* 18 January 1998: 1A, 16A.

[Page 140]
Paul Thomas's comparison of the price he has been getting for wheat in 1999 in comparison with the price his father received in 1953 comes from an article entitled "Farmers Prepare for Low Prices" by Mark Hanson in the *Bismarck Tribune* of 7 February, 1999: A1.

Agricultural economist Won Koo's discussion of farm prices over sixty-five years comes from an article by Patrick Springer, "Consumers, Not Farmers, Benefiting from Ag Research," *Forum* [Fargo-Moorhead] 17 January 2000: A1.

Information on average beef farm incomes for southwestern North Dakota and net farm incomes for the state as a whole come from page 2 of *Heartache in the Heart of the Prairie*, Good Neighbor Project, by Gerald T. Sailer, M.D., Hettinger, North Dakota.

Estimates on the proportion of American workers' income that is spent on food were made by the Ohio Farm Bureau in the spring of 2000. Information on the proportion of income spent by Europeans on food came from Hiram Drache in an interview on August 6, 2000. For underdeveloped countries, the proportion of pay spent on food varies considerably, of course, but Drache suggests that sixty percent is not unusual. In such countries the issue of feeding the people is often not so much the cost of food but the low pay that workers receive.

[Page 143]
Statistics on the declining numbers of farms and farmers have been taken from Dave Danbom, "Cramer Urges Editors to See Reality, Truth," Forum [Fargo-Moorhead] 27 June 1999: E5; Mark Hanson, "State Loses Farmers," *Bismarck Tribune* 27 February 1999; and Genaro C. Armas, "Grand Forks County Tops U.S. in Population Loss," *Grand Forks Herald* 9 March 2000: 1A, 5A. Fargo's population

growth is discussed by Richard Rathge in "Growing Community, Shrinking Labor Pool," *Forum* [Fargo-Moorhead] 1 April 2000: A4. Detailed data on many population factors in rural North Dakota are available at the North Dakota State Data Center at North Dakota State University in Fargo.

The increase in the average age of farmers is discussed in "North Dakota Farmers Aging But Not So Old as National Average," *Agweek* 17 January 2000:5.

Curtis Stofferan is quoted in David Knutson, "Losing the Numbers Game: Census Data Shows Further North Dakota Population Decline from 1990-1999," *Grand Forks Herald* 7 April 2000:1A.

Data on population loss in southwestern North Dakota comes from page 2 of *Heartache in the Heart of the Prairie*, Good Neighbor Project, by Gerald T. Sailer, M.D., at Hettinger, North Dakota.

The problem of rural areas losing political power in the state legislatures is discussed by Richard Rathge in "Rural North Dakota Will Still Command Attention," *Bismarck Tribune* 11 April 2000: 4A and by Jeff Zent in "North Dakota, Minnesota Rural Areas Losing Their Political Clout," Forum [Fargo-Moorhead] 4 April 2000: A1, 8.

Information on the Roman Catholic bishops of Minnesota's stand on government policies that encourage individual ownership of land was taken from "Bishops Seek Assistance for Small Farmers," The Forum [Fargo-Moorhead] 12. February 1999: B2. The North Dakota Catholic bishops have also issued a statement on rural life a brochure entitled *Giving Thanks Through Action: A Statement by the Roman Catholic Bishops of North Dakota on the Crisis in Rural Life*, available from North Dakota Catholic Conference, 227 West Broadway, Suite 2, Bismarck, ND 58501-3797.

In an interview Hiram Drache stated that in Norway the number of farms dropped from 1554,077 in 1969 to 77,461 in 1998 while the size of farms nearly doubled. Sweden has shown a similar change.

[Page 144]

Carol Bly deals with rural life and community in her *Letters from the Country* (New York: Harper & Row, 1981).

Kathleen Norris deals with community factors in times of farm crisis in her *Dakota: A Spiritual Geography* (New York: Ticknor & Fields, 1993), pages 53-55. Other sources on the role of community and types of communities on the Great Plains include Ian Frazier, *Great Plains* (New York: Farrar, Straus, Giroux, 1989); and Richard Critchfield, *Trees, Why Do You Wait?* (Washington, DC: Island Press, 1991).

Programs to help struggling North Dakota farmers make the transition to other kinds of work are discussed in "Program Teaches Farmers Survival Skills," Agweek 22 May 2000: 11.

[Page 147]

The reference to Lee Egerstrom is from page 142 of his *Make No Small Plans: A Cooperative Revival for Rural America* (Rochester MN: Lone Oak Press, 1994), an excellent treatment of cooperatives and their potential for enhancing farming possibilities. See also the report *New Generation Cooperatives* and *The Future of Agriculture* (1999) by William S. Patrie, available from the North Dakota Association of Rural Electric Cooperatives.

[Page 148]

The estimate of increase in U.S. wheat exports that could come with more open trade with China was made by Ron Anderson, "China an Important Wheat Market" Agweek 22 May 2000: 4. Several other articles in that issue of Agweek also give valuable perspective on the issues and possibilities of commodity trade with China.

Information on Middle East markets for sunflower oil was taken from Blake Nicholson, "Middle East a Potential Market for Sunflower Oil," *Agweek* 15 May 2000: 11.

[Page 150]

Information on food security was taken from Lester R. Brown et. al., "Facing Food Insecurity," in S*tate of the World: A Worldwatch Institute Report on Progress Toward a Sustainable Society* (New York: W. W. Norton, 1994): 177-197. Updates can be found in a series of papers in *Vital Signs 2000: The Environmental Trends That Are Shaping Our Future*, ed. Linda Starke (New York: W.W. Norton-Worldwatch Institute, 2000).

The world-wide survey of farmland and its condition is reported in "Global Survey Shows Nearly Half of World's Farmland in Trouble," *Forum* [Fargo-Moorhead] 23 May 2000: 4B.

Richard Critchfield argues in his *Trees, Why Do You Wait?* (Washington DC: Island, 1991) that the welfare of nations depends on their healthy and vital rural communities, especially on pages 9-10, 203-205, and 223-227. Whether or not rural life actually contributes to character or to the strength of a society is a topic of debate among social scientists. For an alternate point of view, see Hiram M. Drache, *The History of Agriculture and Its Relevance to Today* (Danville IL: Interstate Publishers, 1996), pp. 462-467.

The survey Vyzralek cites is found on page 9 of his article.

[Page 153]

Jeff Zent writes about the North Dakota State University research and the USDA Baseline Projections on possible improvement of commodity prices in "Study Predicts Brighter Future for U.S. Farmers," *Forum* [Fargo-Moorhead] 7 June 2000: A1, A14.

Jack Zaleski's editorial "Change Is Not Always a Good Thing" appeared in the *Forum* [Fargo-Moorhead] 13 February 2000: E5.

The future looks good for economic returns from people interested in visiting natural environments for their leisure activities because North Dakota has maintained the beauty of its landscape. Tourism is the fastest-growing industry in the West and the largest private employer in seven of the eleven western states.

North Dakotans are proud of the accomplishments of their sons and daughters. This monument to Brad Gjermundson, four-time PRCA world saddle bronc champion, sits atop a hill overlooking Marshall, in Dunn County, where he grew up.

Home on the Range

Oh, give me a home where the buffalo roam,
Where the deer and the antelope play;
Where seldom is heard a discouraging word,
And the skies are not cloudy all day.

Where air is so pure and the zephyrs so free,
And the breezes so balmy and light;
That I would not exchange my home on the range,
For all of the cities so bright.

How often at night, when the heavens are bright,
With the light from the glittering stars,
Have I stood there amazed and asked, as I gazed,
If their glory exceeds that of ours?

Refrain:
Home, home on the range,
Where the deer and the antelope play;
Where seldom is heard a discouraging word,
And the skies are not cloudy all day.

- Traditional cowboy song

North Dakota is poised to take advantage of new opportunities the present technological revolution is making available today.
Hope lies in our productive landscape, like this scene from Grand Forks County in the fertile Red River Valley, which helps
create a renewable product – quality food – each growing season.

Opportunities to come

Edward T. Schafer
Governor of North Dakota
1993-2001

I took my cabinet and staff to Renville County for our Capital for a Day program in June 1999 and it was an eye-opener. We saw field after unplanted field, looked at the shocking photos of the spring floods on Ali Bahl's web site, and heard about the difficulties facing businesses in Mohall. But we also visited a local farmer who had diversified into canola, was practicing no-till cultivation and was trying something new - fall seeding of dormant canola seed.

Like wild swings in the prairie weather, agriculture in North Dakota is a study in extremes.

With the recent tough times, there are lots of expressions of frustration, anger and resignation. Farmers and ranchers wonder if it makes sense to go on. These are all understandable, sensible reactions to low commodity prices, seven years of disaster and repeated outbreaks of crop disease. Such extremes make for a powerful argument for adequate risk management programs.

At the same time, I hear optimism and hopeful comments. Farmers tell me about the changes they made to their operations to improve profitability. I'm given hope by someone who pulls me aside and says, "Hey, Ed, I've found a way to make some money."

We cannot ignore the hardships. They have a real impact on our friends and family members, local businesses and the state's economy.

But we also cannot let hardship define agriculture in North Dakota. Searching for, and creating new opportunities, should be our focus. If we keep that focus at the forefront, then farming, ranching and all the related agribusinesses can prosper and retain their leading place in the state's economy and way of life.

I have been a big proponent of opening export markets as necessary for making more money in North Dakota agriculture. If we can get the Chinese to start eating products made with North Dakota crops; break down the sanctions against countries like Cuba, Iran and Libya; and pursue an aggressive trade strategy, then opportunities for our farmers will grow.

Much of this is up to the federal government. The role of the governor and others like me in state government is to be a loud voice, shouting at the slow movers in Washington to get on with it. That's true whether its trade policy, crops insurance reform, or easing regulations.

At the state level, I have seen my job as creating opportunity. One of my priorities has been to increase spending on agriculture research, and we'll see a payoff in the years to come. We have encouraged irrigation projects allowing farmers to expand into high-quality vegetables and more uniform commodities. Once we created a better legal framework for investment, North Dakota also made great strides in value-added processing, a subject addressed elsewhere in this book.

Perhaps more than anything, technology also holds the key to the future of agriculture in our state. I am especially optimistic about the impact of genetically modified crops. Two decades from now you might see a farmer in northeastern North Dakota growing a genetically modified flax seed that produces cooking oil doctors prescribe to clean clogged arteries. Next to that field will be another one of hemp, bred to grow with blue leaves, so no one mistakes it for an illegal drug. There may be a farmer in southeastern North Dakota twenty years from now whose soybeans produce a chemical used in our manned mission to Mars. An organic farmer in the west might succeed by responding to the Germans' insatiable demand for sunflower-flavored breads containing no pesticides. These kinds of crops will demand high prices, specialized knowledge and skilled farming techniques. Our farmers have shown themselves able to adapt to such demands.

Sounds like a dream, a little too futuristic? Maybe. But two decades ago, could anyone imagine North Dakota farmers using global position satellites to make decisions about where to plant or fertilize? Or turning to genetically modified soybeans to cut their pesticide costs? Or farmers growing profitable crops of potatoes in central North Dakota? Just as technology will provide opportunity for North Dakota's farmers, it will also provide opportunity for our rural communities. As I said in my final State of the State address in January 2000, agriculture will remain a powerful economic engine.

North Dakota is poised to take advantage of the vast opportunities the technological revolution is making available to every willing participant. Classrooms will become global. Teachers will pull information and even people from throughout the world into their classrooms. Corporate executives will be as successful in Watford City as on Wall Street.

When high school students today are at the peak of their career, they will use computers that are seventy million times more powerful than mine and will communicate with one hundred percent natural human language and emotion. Entrepreneurs, local business owners, international firms paying our highly educated workers — with technology, all of these can strengthen the rural economy, working in tandem with agriculture to make sure North Dakota's future shines.

Pioneers broke the sod to build the state of North Dakota and I have no doubt they would regard our state today as a marvel or a miracle. If we keep faith with their hard work and their belief in the future, and use the opportunities at hand, North Dakota will remain that miracle for generations to come. ∎

"LIKE I SAY, WE'LL TRY THIS ONE MORE TIME... AND THEN HOPE AND PRAY THERE'S SOME LIGHT AT THE END OF THIS TUNNEL."

By Trygve Olson. Used by permission of The Forum, Fargo-Moorhead.